TECTONIC STRESS IN THE LITHOSPHERE

PROCEEDINGS OF
A ROYAL SOCIETY DISCUSSION MEETING
HELD ON 10 AND 11 APRIL 1991

ORGANIZED AND EDITED BY
R. B. WHITMARSH, M. H. P. BOTT,
J. D. FAIRHEAD AND N. J. KUSZNIR

LONDON
THE ROYAL SOCIETY
1991

Printed in Great Britain for the Royal Society
by the
University Press, Cambridge

Book: ISBN 0 85403 448 X
Journal: ISBN 0080 4614

First published in *Philosophical Transactions of the Royal Society of London*,
series A, volume 337 (no. 1645), pages 1–194

The text paper used in this publication is alkaline sized with a coating which is predominantly calcium carbonate. The resultant surface pH is in excess of 7.5, which gives maximum practical permanence.

British Library Cataloguing-in-Publication Data

A CIP catalogue record for this
book is available from
the British Library

ISBN 0-85403-448-X

Published by the Royal Society
6 Carlton House Terrace, London SW1Y 5AG

CONTENTS

The nature and determination of stress in the accessible lithosphere

By T. R. Harper[1] and J. S. Szymanski[2]

[1] *BP International Limited, Sunbury Research Centre, Chertsey Road, Sunbury-on-Thames, Middlesex TW16 7LN, U.K.*

[2] *Department of Energy, PO Box 98608, Las Vegas, Nevada 89193–8608, U.S.A.*

Results of stress determination show that stress state is characteristically heterogeneous and apparently unpredictable. Characterization of *in situ* stress state requires the determination of stress at individual locations and then spatial if not also temporal extrapolation. A variety of different measurements are used as a basis to determine stress state. Thus an adequate understanding of both the nature and origin of rock stress is essential. Representation of the lithosphere as a non-equilibrium, dissipative, dynamical system is shown to be consistent with observations of stress state and fluctuation of crustal displacement. The evolution of rock cores subject to a variety of perturbations in the laboratory can similarly be shown to be consistent with the evolution of a dissipative, dynamical system, driven in part by stored strain energy. These observations are inconsistent with the assumption that rock stress can be adequately represented by superposed traction and internally balanced stresses arising from quasi-static processes. The potential value of analyses of the dynamics of rock stress evolution is emphasized as a means to simplify the apparent complexity arising from present perceptions.

1. Introduction

The motivation to characterize the distribution of forces in the lithosphere stems from geological hazards, engineering activities and resource exploration. Because of the coupled nature of geological processes, these forces profoundly influence not only the capacity for work, but also a wide range of transport phenomena, for example fluid flow.

For reasons of cost and practical difficulty, determinations of stress are often both incomplete and sparse. An ability to characterize the distribution of stress throughout a given volume of rock on the basis of a very limited suite of measurements is therefore essential. A predictive capability based on extrapolation from point measurements to neighbouring volumes would be very valuable.

The intent of this paper is to present a perspective of current practice and perceptions and a basis for future developments in the characterization of stress state from individual measurements. Inevitably, this involves appraisal of the nature and genesis of stress, being a logical basis for interpreting individual measurements and extrapolating from a sparse data set. We focus on the accessible lithosphere, by which we mean that part of the lithosphere to which we can potentially gain access for measurement purposes. In practice, this is usually less than 5 km, but here we

Phil. Trans. R. Soc. Lond. A (1991) **337**, 5–24
Printed in Great Britain

assume a depth limit of 10 km. Other than for purposes of illustration, we make no attempt to document previous determinations or characterizations of stress.

With the exception of the term residual stress in place of initial stress, the terminology of Timoshenko & Goodier (1982) is used. The adjective dynamical is used to describe systems in which the configuration at a later time is a function only of the configuration at an earlier time. Taking an example from Arrowsmith & Place (1982) as an illustration, if the state of a system is given by the vector $\boldsymbol{x}(t)$ at time t then the time-development of states is governed by the dynamical equations (or equations of motion) $\dot{\boldsymbol{x}} = \boldsymbol{X}(\boldsymbol{x})$.

2. Current practice and perspective

(*a*) *The nature and origin of stress*

Early in the history of rock stress determination, it was recognized that stress differences are characteristic of *in situ* stress state (Hast 1967). Hast noted differences between all three principal stresses at a point. In 1978, following a review of a large number of stress determinations, McGarr & Gay concluded that it is not possible to generalize concerning the magnitudes of horizontal stresses except that they usually increase with depth. They may be lesser or greater than the vertical stress. Indeed, the vertical stress may be less than or greater than that of the calculated lithostatic value (McGarr & Gay 1978; Barton 1982; Martin 1990). McGarr & Gay (1978) concluded that the orientation of horizontal stress may be either relatively homogeneous, with local anomalies, or completely incoherent from site to site. The same authors concluded that no reference state of stress exists that represents any particular region and also has worldwide applicability. The degree of consistency of the direction of principal stresses projected onto the horizontal plane is evident primarily at large scales, notably those which are a significant fraction of plate dimensions. However, bearing in mind that these directional data provide extremely limited information concerning the stress tensor, the inevitable conclusion is that stress state in the upper lithosphere is characteristically heterogeneous. (Apparent exceptions may occur locally in weak rock unable to support a shear stress, e.g. in highly overpressured formations.) Because stress state is unpredictable (McGarr & Gay 1978; Hiltscher *et al.* 1979), it is necessary to perform measurements to determine the stress state in any given volume of rock.

A summary of the origin of stresses in the lithosphere is given by Ranalli (1986). Forces promoting translation in the plane of the plate are body forces acting in the vicinity of divergent and convergent margins. Those resisting motion are surface forces including mantle drag, transform fault resistance, frictional resistance associated with collision, viscous resistance of the mantle of slab sinking and bending forces in the vicinity of trenches. Intraplate stress is strongly influenced by gravitational forces. All these forces maintain a reservoir of strain energy in the lithosphere which is relieved at approximately the same rate as the rate of input (Bott & Kusznir 1984). These authors note that a concentration of stress in the upper lithosphere can be inferred from the variation of rheology with depth.

All these forces are normally viewed as quasi-static, the influence of the dynamics of change and characteristics of the dynamics, such as fluctuation, having received little or no attention.

Many investigations requiring determination of stress state are concerned with volumes of rock comprising a very small part of a tectonic plate. Many are limited

Table 1. *Bases for stress determination*

basis	measurement variable
earthquake	seismic radiation
hydraulic fracture	fluid pressure
borehole	geometry of finite deformation, reloading strains
core	relief strains, evolution of relief strains, reloading strains, finite deformation state

to the accessible lithosphere. At this much smaller scale, rock stresses are also typically treated as resulting from external forces (surface forces and body forces) alone. However, it has long been recognized that strain relief measurements in isolated bodies of rock, free of surface forces, characteristically reveal the presence of residual strains (Emery 1963). Consideration of the possible modes of genesis (Voight & St Pierre 1974; Holzhausen & Johnson 1979), suggests that residual stresses are characteristic of the majority of rocks derived from the accessible lithosphere. However, with few exceptions, they have been ignored. This might be acceptable if the scale of balanced domains is much larger than the volume of rock of interest or the influence of residual strain and stress are otherwise insignificant. If this is not the case, as the strain relief measurements and genetic considerations suggest, then treatments of stress in terms of surface and body forces alone may be misleading.

(*b*) *Determination of stress at a point location*

Table 1 presents a simple grouping of the principal methods of stress determination. Although the distinction between different methods may not be as straightforward as suggested by the table, it serves to emphasize that the measurement basis is not common to all methods. If the results of different methods are to be comparable (i.e. quantitatively relatable to the single parameter stress), this implies a knowledge of the relationship between wave propagation from earthquakes responding to a long-term strain energy accumulation; the opening and closing pressures of a fracture; almost instantaneous strain relief; time-dependent strain relief; and, progressive failure. All of these measurements are interpreted in terms of tractions (external forces). The question of whether we have adequate understanding to compare the results of different test methods will be considered in §3 (apart from that relating to focal plane analyses). It is also apparent that the four bases of stress determination (table 1) represent three length scales extending to approximately three orders of magnitude. For results to be comparable implies either a scale-independence of stress state or a knowledge of any scale-dependence.

For many geological and engineering purposes, stress is determined using hydraulic fractures, boreholes or rock core. Theoretically, these methods all have the potential for complete determination of the stress tensor from a set of measurements at a point location but in practice this only applies when rock coring is undertaken.

The creation and re-pressurization of hydraulic fractures is a well-tried and widely accepted technique of stress determination. Combined with a technique for identifying fracture orientation, multiple boreholes are normally required to determine the stress tensor. The conventional method is limited to sections of borehole devoid of natural fractures but Cornet (1982) developed a technique specifically based on the re-pressurization of mutually inclined natural fractures.

Major advantages of the hydraulic fracturing technique are the capability to operate at most drilled depths, the relative simplicity of the measurement principle and the considerable body of experience gained to date. The fluid pressure required to open and close the created fracture, assumed to be a mode 1 fracture, can be measured. (See Hickman & Zoback (1982) for a more detailed account.) In contrast to the majority of other stress determination methods, measured pressure is directly equatable to a traction representing one component of the stress tensor. Analysis of fracture initiation pressure in previously unfractured borehole additionally allows interpretation of the maximum traction acting normal to the borehole axis.

Borehole deformation, or breakout, is now widely used to identify the directions of horizontal stress. Caliper logs are often run for reasons other than stress determination, making the method inexpensive as well as practical. In contrast to hydraulic fracturing, the method relies entirely on inelastic processes (failure of the borehole wall). Determination of stress magnitude would require an ability to relate the finite change of borehole wall geometry to the perturbed stress state around the borehole by means of a quantified knowledge of the progressive failure process. Despite research efforts, the potential for determination of stress magnitudes from breakout logs is unpromising, in part because of the ill-defined complications of mud chemistry, and undocumented transient mud pressures and temperatures. Indeed, based on numerical simulations, Zheng *et al.* (1988) concluded that breakout geometry is both stress history-dependent and non-uniquely related to *in situ* stress. Other techniques based on borehole deformation are less used. Some involve reloading of the borehole wall. Other, technically attractive research techniques are based on sidewall coring (Schmitt *et al.* 1985). No attempt is made here to comprehensively review all techniques.

Determination of stress based on rock core, initially confined to overcoring techniques (Hooker & Bickel 1974; Leeman 1969), has been extended to a range of other core-based methods (Teufel 1981). Use of overcoring methods, although able to determine both magnitude and direction of stress and even the complete stress tensor in a single determination (Leeman & Hayes 1966), has been constrained by depth limitations. The greatest depth capability recorded, 500 m (Hallbjorn *et al.* 1990), is substantially less than that of many oil field and geothermal operations. The principle of overcoring is based on the strain induced when a core is created, i.e. when the volume of rock which becomes the core is relieved of the applied stress. This strain relief process is assumed to be elastic (reversible) so that the core may be subsequently reloaded to determine suitable moduli from which to relate strain relief to stress relief. However, another technique (anelastic strain relief (Teufel 1981)) is entirely reliant on the time-dependency of core strain evolution. No allowance is made by the overcoring technique for this characteristic time-dependency. This would only be serious if the deformation properties of the core are irreversible and significantly different to *in situ* properties before overcoring, or if significant inelastic strains develop during a test (assuming that representation of time-dependent behaviour is not achieved by additional testing).

Overcoring strain measurements, in common with hydraulic fracturing pressure measurements, are interpreted as *in situ* tractions. Therefore, no allowance is made for any irreversible change of internal forces. One notable exception is the work of Hiltscher *et al.* (1979). These authors reported residual stresses in free boulders of 1–15 MPa, comparable with *in situ* stress magnitudes typically less than 30 MPa.

Attempts to deduce stress magnitudes from time-dependent strains observed when core is brought to the surface are also reported only in terms of tractions (Teufel 1985) and the potential influence of internal strains, inferred to be present from experiments involving residual strain release, are again not taken into account. It could be argued that the residual strains are solely a result of the unloading to a traction-free condition and cooling to ambient conditions and not present *in situ*. However, not only would this be unsubstantiated, but consideration of the genesis of residual strains in the context of well-known geological processes including changes of depth of burial (change of confining pressure and temperature), cementation by diagenetic fluids, pressure solution, and fracturing, strongly suggests that this is unlikely.

Use of differential strain curve analysis for stress determination (Strickland & Ren 1980) is based upon hydrostatic reloading of cubes cut from core. Stress directions are inferred from directional differences of reloading strains. Attempts have been made to deduce stress magnitudes from changes of slope of the stress–strain curves, inferred to correspond to the closure of microcracks formed during core strain relief. It is thus assumed that strain relief is essentially reversible, that the effect of a lack of shear stresses and other effects during core reloading are insignificant, that core dilation relates only to microcrack opening and that the effect of creating new free surfaces and the shape of the cube are insignificant. Most, if not all, of these assumptions have not been justified.

This necessarily brief review of methods of stress determination is neither comprehensive nor exhaustive. We have deliberately avoided a focus on details of test procedure. Readers wishing to learn either more of each of the techniques mentioned or of alternative techniques not mentioned here are strongly urged to examine the independent literature. Our purpose in this scant review is to highlight some of the potentially important assumptions and apparent inconsistencies in order to assess the present ability to determine rock stress state at a point location. First, we ask whether the fundamental nature of rock stress is adequately understood. Because the role of internal forces and their relationship to surface forces has not been defined, the essentially universal tendency to ignore internal strains remains, at best, unjustified. At worst, we must conclude that the nature of rock stress is not adequately understood. Second, in view of the fact that the various test techniques do not yield the same information, is there an adequate basis for comparison of the results achieved by different methods? If the role of internal forces is significant, or if there exists an undefined scale-dependency within the length scale spanned by the various forms of measurement, then the answer to this question is negative. Indeed, if a given test is influenced by internal forces in an ill-defined manner, even comparisons of the results of the same test conducted on different rock types may be misleading.

(*c*) *Characterization of stress distribution*

Given a group of individual stress determinations it is necessary to interpolate and extrapolate these in space, and sometimes in time, to characterize the stress distribution. Viewing the literature as a whole, we note a pronounced tendency to assume a uniform 'regional' or 'far-field' stress state (tractions) which is derived by assuming that directional data may be averaged and that depth variation may be reduced to a linear increase of each stress component. The linear increase with depth is either explicitly or implicitly related to gravitational body force or frictional criteria. Any departure from this uniformity may be either ignored, or related to local structure or contrasts of present-day material properties responding to 'far-

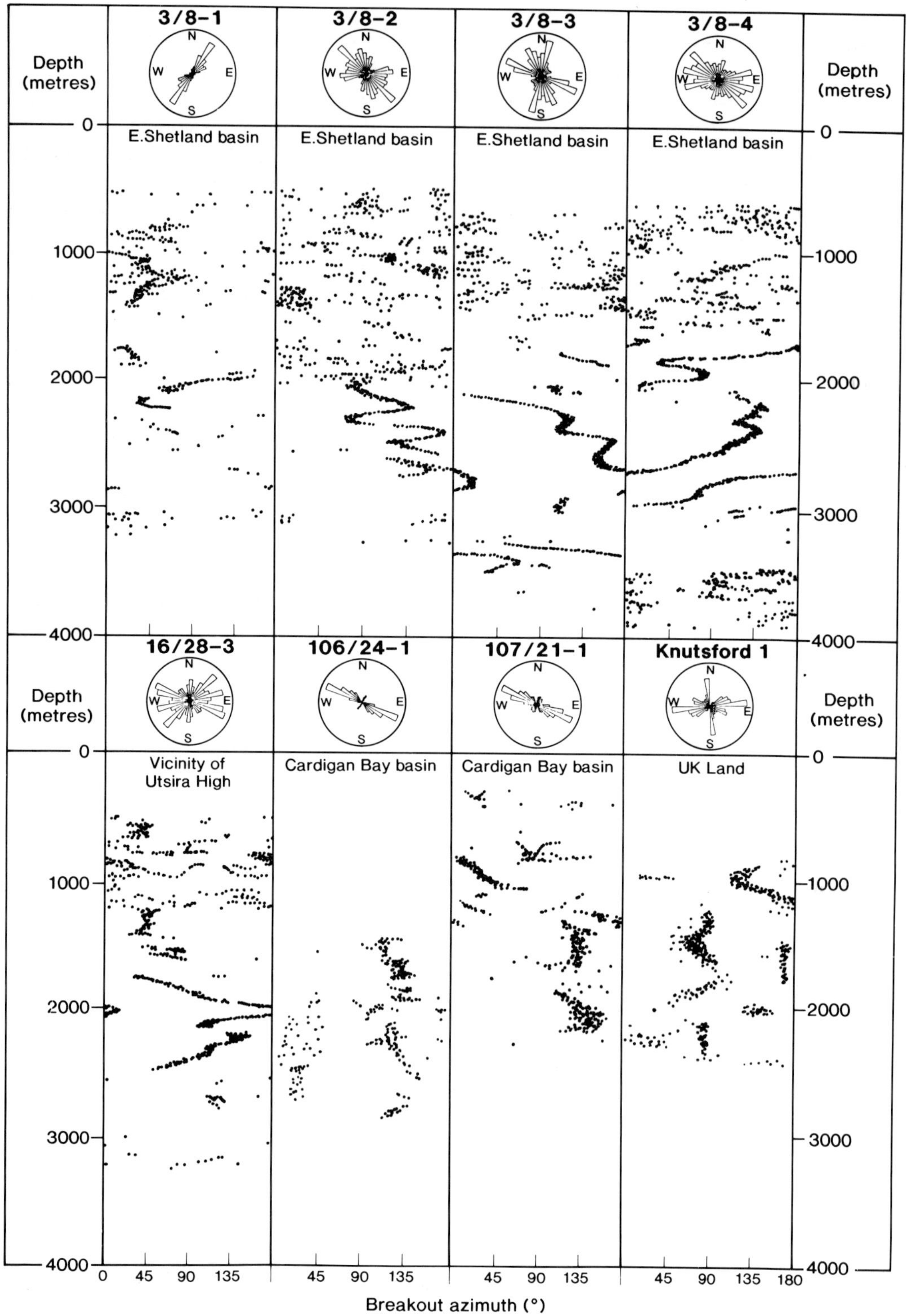

Figure 1. For description see opposite.

field' tractions. The distribution of stress in time is normally ignored and therefore implicitly assumed to be unimportant.

Many records demonstrate that averaging of directional data can be misleading and helpful only when applied to localized volumes of rock with a demonstrated consistency of stress direction or if the variability concealed by the averaging process is of no consequence to the characterization of stress state (McGarr & Gay 1978; Martin 1990; Shamir *et al.* 1990; Dames & Moore 1978; Hiltscher *et al.* 1979, fig. 1). Often the assumption that averaging is justified is only implicitly stated.

Stress contrasts may be present across major stratigraphic discontinuities (Haimson & Lee 1980), and stress gradients and stress contrasts can be associated with faults, whether or not these are recognizably active (Anderson *et al.* 1983; Harper & Szymanski 1981*a*; Martin 1990). As noted by Hickman *et al.* (1988) spatial fluctuations in stress magnitude and orientation have been documented in homogeneous granitic rock. Kry & Gronseth (1982) concluded that the detailed nature of the *in situ* stress state in the Deep Basin of Alberta and British Columbia cannot simply be related to lithology. This conclusion is also demonstrated by the results compiled by Szymanski (1979) for tuff and crystalline rock in the Great Basin and reported by Dames & Moore (1978) for sediments in the Northern Appalachian Basin. These observations demonstrate that stress is not solely a function of lithology, present-day structure, and conventional rock mechanical properties.

Indeed Hast (1967), Voight & St Pierre (1974), Harper & Szymanski (1981*b*) and Szymanski (1989) and others have emphasized the time-dependency of stress. A combination of the object of the investigation and the rate and nature of energy transfer in the lithosphere determine the timescale of stress changes which may be relevant. For example, analysis of stress changes occurring during periods of only hundreds or thousands of years would be warranted for assessment of an active fault hazard, or for safety analyses of candidate high level waste repositories, particularly those sited in areas subject to high rates of energy transfer. Szymanski (1989) concluded that the rate of stress evolution at Yucca Mountain, Nevada, is sufficient to potentially influence such a facility. By numerically simulating the influence of a change of 'far-field' stress state on a jointed but otherwise uniform rock mass, Lemos *et al.* (1985) emphasized the influence of stress history on non-uniformity of stress distribution. These reports suggest that an understanding of the evolution of stress may be an essential prerequisite to an adequate interpretation of stress state. Static models may be insufficient.

3. A nonlinear systems perspective

(*a*) *Introduction*

The preceding review suggests that current conceptualizations of both the nature and origin of stress provide an incomplete basis for the characterization of stress

---

Figure 1. Examples of the variation of direction of maximum horizontal stress with depth inferred from borehole breakout records in the North Sea, Cardigan Basin and onshore U.K. All boreholes are deviated less than 5° and defined by an ellipticity threshold of 0.05 in (*ca.* 13 mm). (Note that borehole ellipticity can be influenced by factors other than stress. The potential influences of bottom hole assembly stiffness contrasts, logging tool rotation, steeply dipping thin beds of contrasting strength and the effect of varying well trajectory have been assessed in interpretation of the pronounced progressive depth variation encountered in the E. Shetland wells. Only the effect of well trajectory was not eliminated (for lack of data) but since the deviation from the vertical is so slight, it is extremely unlikely that well trajectory effects are significant.)

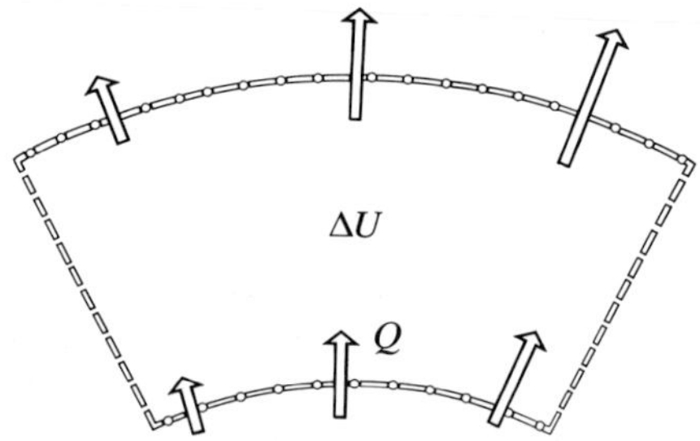

Figure 2. Conceptual thermodynamic model of the lithosphere. ⊐ ⊏, Mechanical work (time- and space-dependent), W; ⊐○⊏, mechanical work and heat flux (both time- and space-dependent). $\Delta U = Q - W$ is the change of internal energy. Note that the boundaries of the conceptual model are fixed in space and time.

state. Ideally, we would be better able to relate the different measurements which provide the basis for individual stress determinations. As a minimum, we must be able to relate the process which yields the measurement, the values represented as stress which are computed from this measurement and the engineering or geological process of interest. To translate these individual observations into a characterization of stress distribution, it is sometimes required to explicitly address the evolution of stress as a function of time as well as space. Unless we can be confident that stress distribution is independent of time, that is, independent of geological history and dominated by present-day quasi-static surface and body forces, it is an implicit requirement to consider stress evolution.

(*b*) *A conceptual thermodynamic model*

Figure 2 represents a generalized conceptual model of a section of the lithosphere. It is convenient to consider all of the lithosphere, but the model could equally well be confined to the accessible lithosphere. The model is based upon energy transfer and transformation (i.e. it is thermodynamic). A flux of energy and matter occurs across all boundaries, varying as a function of both time and space. For simplicity, the radial boundaries are assumed to be adiabatic, otherwise energy transfer across boundaries is by both work and heat transfer. At the lower boundary, energy is transferred from the mantle by time- and space-variable heat transfer or by work associated with shear and normal stresses acting on the base of the lithosphere. Mantle drag forces vary spatiotemporally in part because the profile of the base of the lithosphere is irregular (Wan *et al.* 1989) and varies with time, but this is not a necessary condition for this spatiotemporal variability of energy transfer. There is no reason to assume that forces acting on the lateral boundaries, promoting circumferential plate translation, are time-invariant, because they are associated with Rayleigh convection in the mantle and the lithosphere deforms internally in a time-dependent manner. Incessant heat flux and redistribution of matter occur at the Earth's surface. The latter, in conjunction with density changes caused by phase changes both in the mantle and in the lithosphere in the vicinity of the aesthenosphere–lithosphere boundary, and by other effects, results in changes of gravitational body force. The lithosphere is thus perceived as a non-equilibrium (in the thermodynamic sense), open system.

Energy is transferred within the system. In common with most natural systems heat generation accompanies work, whether mechanical, nuclear, electrical or other, and by mechanisms such as friction and thermoelastic effects, particle bombardment or current flow. From the second law of thermodynamics, we note that the system is dissipative. If $X_1, \ldots, X_n$ denote a complete set of macroscopic variables of such a system, their time evolution will take the form (Nicolis & Prigogine 1989).

$$\partial X_i/\partial t = F_i(X_1, \ldots, X_n, \boldsymbol{r}, t),$$

in which F_i may be complicated functions of the Xs and their space derivatives, as well as explicit functions of space $\boldsymbol{r}$ and time t. This evolution equation implies that a variability in time and space, with associated gradients, is characteristic of lithospheric systems for reasons additional to variable boundary flux.

The energy transformation and transport phenomena occurring within the system are characteristically nonlinear. Feedback loops occur in processes such as flow and chemical reaction (Nicolis & Prigogine 1990) and strain localization (Cundall 1990). Coupled processes with feedback relationships are widespread, such as fracturing and fluid pressure change, variation of rheology with strain rate and temperature, Rayleigh convection, diagenesis and flow of mineralizing solutions.

A remarkable number of useful conclusions can be drawn from this simple conceptual model. First, the model suggests that gradients are a characteristic of the lithosphere. This is consistent with the observations of stress heterogeneity (§2) and with the concept of internal forces. Second, the genesis of stress (or strain energy) occurs in the context of an evolving system; treatment of stress as an isolated variable or by superposition, such as linear thermoelasticity, may be unhelpful or misleading. Third, as a nonlinear, open, dissipative dynamical system, at least two very influential characteristics should be expected of the lithosphere: fluctuation and self-organization (Nicolis & Prigogine 1989). It seems that many geoscientists have traditionally viewed geological processes as basically monotonic. Quite in contrast, the simple dynamical model discussed here suggests that fluctuation, perhaps superimposed upon a monotonic trend, is incessant. The existence and perhaps surprisingly high rates of fluctuations are demonstrated by vertical crustal movements deduced from geodetic levelling (Whitten *et al.* 1977), horizontal strain measurements (Smith & Kind 1972), and by geomorphologic, stratigraphic and structural (Grant 1990) records. The work done during these fluctuations, conventionally neglected, can be expressed as a coherent, unidirectional effect such as heat transfer (Kurzweg 1985) and mass transfer (Chatwin 1975; Nilson & Lie 1990). Moreover, the conceptual model suggests that treatment of stress genesis as a quasi-static process (Warpinski 1986; Prats 1981) has limited potential.

Recognition of the truly thermodynamic nature of geological systems implies the daunting task of analysing the many coupled processes. However, one characteristic of dissipative systems offers at least the hope of remarkable simplification of the task of characterizing stress distribution. In contrast to conservative systems, dissipative systems are capable of eliminating the effect of perturbations and returning to a reference or attractor state. (They may also amplify small perturbations.) Fractal attractor states have been reported for a variety of geological features (Turcotte 1989; Main *et al.* 1990; Velde *et al.* 1990). The concept of self-organized criticality (Bak *et al.* 1988), having power law spatial and temporal correlations, has been applied to earthquakes (Bak & Tang 1989). Keilis-Borok (1990), recognizing that quantitative description of all the nonlinear, coupled relationships governing lithosphere behaviour is unrealistic, has emphasized the need for a generalized description. With the exception of Sornette *et al.* (1990), these concepts have yet to be applied specifically to stress evolution.

(c) *Laboratory tests*

(i) *Introduction*

The recognition that geological processes occur within the context of a system (Harper & Szymanski 1981*b*), such as outlined in §3, implicitly questions the applicability of laboratory tests staged such that many variables are constrained to be constant. This severely reduces or eliminates the interdependence characteristic of geological systems. Nevertheless, they are helpful providing that the effect of the imposed experimental conditions is fully recognized.

Laboratory tests can be used to indicate the potential significance of internal forces. In addition, they can be used to show that simple superposition of forces corresponding to residual stresses (stresses which exist in the presence of traction-free boundaries) and the hypothetical forces instantaneously arising within the material from the static application of tractions, cannot be used to describe the internal force distribution of rock *in situ*. The phrase interaction forces is therefore introduced here as a category of internal forces. It is used to describe the force distribution which occurs for given mechanical and thermal boundary conditions at a given time in the evolution of the rock, minus those forces which would result from a hypothetical static response to the application of the prevailing tractions (and thermal boundary conditions) if the rock behaved as a conventional continuum. This is certainly not a rigorous description. Moreover, it is of limited helpfulness because the absolute nature of such stresses cannot be quantitatively determined by experimentation. We shall see that it is very probable that even the 'residual' stress state in rock cannot be uniquely determined because a perturbation imposed to investigate residual stress causes change. However, interaction forces, although a loose description, at least refer to *in situ* geological conditions where tractions would normally form part of any description of stress state. The term is intended to emphasize the critical role of the evolving interactions which occur in a system of particles (grains, crystals, mineral groupings) or particles and cement. (Evolving interactions between much larger volumes of material are also implied, such as strata of differing properties or blocks bounded by discontinuities.) This new terminology is necessary because none of the established terms appear to be appropriate. From the subsequent review of laboratory experiments it may be inferred that the internal force distribution in many rocks evolves irreversibly when subject to perturbations such as the application or removal of external forces or temperature change. Consequently, neither initial stresses nor residual stresses, which represent a condition of zero external force, are identical to that component of the internal force distribution under load which does not simply reflect transmission of the external forces. Nor are these interaction forces solely self-balancing in the presence of external forces so the term self-equilibrating is also inappropriate.

(ii) *Observations*

The evolution of interaction forces and their relationship to surface forces, or tractions, can be inferred by subjecting suitable rock core, such as a sandstone, to a range of perturbations. Acoustic emission during creep experiments implies irreversible changes involving interaction force redistribution. Microseismicity occurs not only during loading but also during creep recovery (Harper 1970). That is, application of surface forces influences interaction forces. This is consistent with the torsion loading experiment conducted on granite by Tan & Kang (1980), who, after

Figure 3

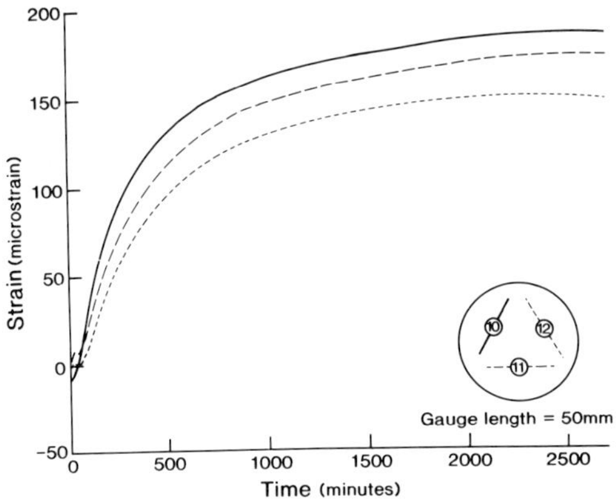

Figure 4

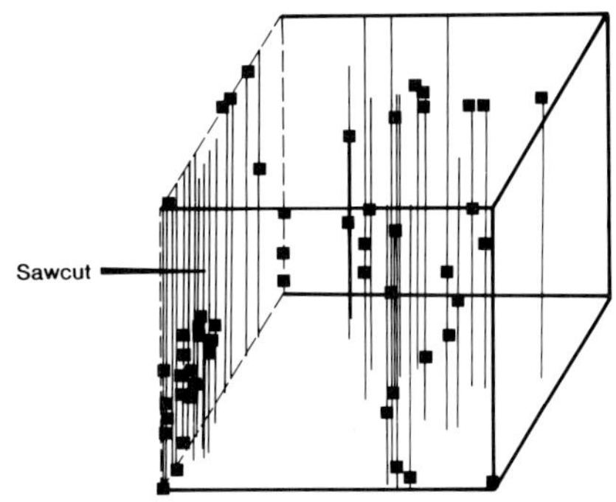

Figure 3. Time-dependent strain relief of Permian sandstone cored at a depth of 3042.4 m from vertical well 42/30-7. Strains were induced in the 100 mm diameter core of original length 150 mm by a sawcut normal to the core axis and 60 mm from the strain gauges placed on the flat surface normal to the core axis. (Note that the magnitudes of core strains induced by the single sawcut represent the upper limit of the spectrum of magnitudes observed in the Lower Leman sandstone in this area.)

Figure 4. Approximate distribution of acoustic emission accumulated in a core of Lower Leman sandstone during a period of 10.35 h by a sawcut normal to the core axis. The distribution of events is approximated by recording only those events detectable at all six transducers. The cylindrical core is represented as a cube. Each event is denoted by a rectangular box with a vertical line to indicate the relationship to the boundaries of the core.

unloading, observed a finite angular strain opposite in sense to the applied torsional load. Records of the release of internal strain energy to generate surface forces are rare. However, Price (1966) reported a time-dependent progression of extensional strain during the first loading stage of a compressional creep experiment. This implies that, for a period greater than 50 days, the nodular, muddy limestone sample was exerting a stress in excess of that applied (approximately 36 MPa). This rare example suggests that internal forces can influence surface forces. It is well known that changing sample geometry (cutting a free surface) results in strain and displacement of the sample boundaries, some examples of which are noted by Tullis (1977). Figure 3 records the time-dependent strain relief observed by strain-gauging one end of a North Sea sandstone reservoir core and sawing a slice of material from the other end. Successive cuts of this type have been found to initiate successive periods of strain relief, the strains in each case being extensional and appearing to stabilize in a period of approximately half a day. Acoustic emission is observed throughout the core, not simply in the vicinity of the saw cut (figure 4). Strains have been observed at a distance more than 10 times the largest dimension of the created free surface (figure 5), revealing that measurable affects can in some tests be observed for very small geometry changes and that perturbations may propagate to distances substantially greater than can be explained by elastic behaviour. Displacements of sample surfaces, accompanying irreversible sample strains, can also be induced by imposing minor temperature changes. Figure 6 shows the response of reservoir sandstone, revealing that finite extensional strains develop in response to cyclic minor temperature change. Again, imposition of successive temperature cycles results in successive increments of time-dependent extensional strain. At the

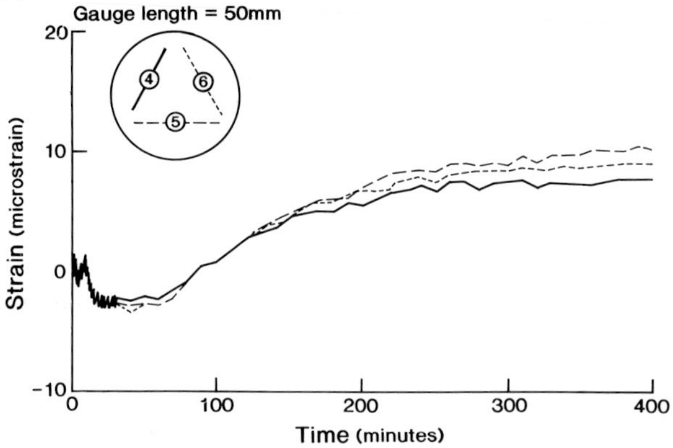

Figure 5. Strain relief remote from perturbation. Strains observed 235 mm from the base of an 8 mm diameter, 20 mm deep undercore in Lower Leman sandstone cored at 3031.42–3031.67 m in vertical well 48/6-30, North Sea. The strains were measured in a plane normal to the core axis.

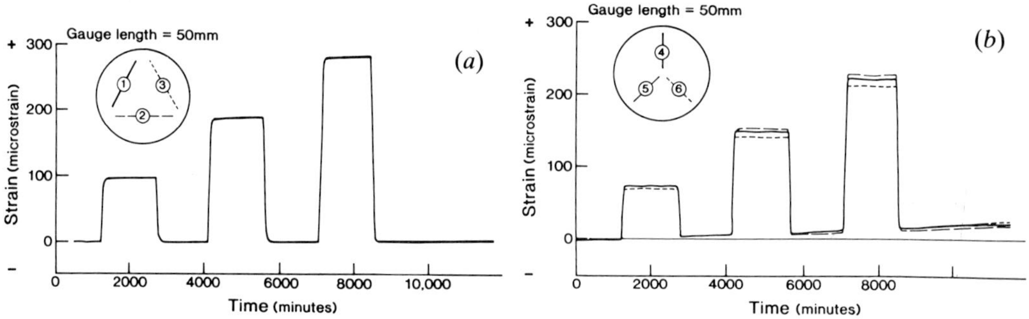

Figure 6. Response of sandstone to incremental heating cycles. Carboniferous sandstone cored at 3143 m from North Sea well 42/30-7. The 100 mm diameter core sample was strain-gauged normal to the core axis and heated with a steel control sample in a temperature-controlled oil bath.

conditions of zero confining pressure pertaining in the tests reported in figures 3–6, a distinct asymmetry is evident in that the finite strains developed in a time-dependent manner after returning clean sandstone cores to the reference state are always extensional.

(iii) *Interpretation*

Although more experiments of a different nature are required for a more convincing demonstration, these experiments strongly infer that surface forces (and/or displacements) and interaction forces (and/or displacements) are interdependent. The remote acoustic emissions and strains occurring in response to a local geometric perturbation reveal that the whole core may evolve. The effect of a local perturbation is propagated in a time-dependent irreversible manner away from the perturbation. A consequence of the local perturbation is a finite displacement at the sample boundaries. Contrary to the conclusions drawn by Holzhausen & Johnson (1979), but consistent with the hypothesis of Voight & St Pierre (1974), it is concluded that microscopic complexity, approaching apparent disorder, can be converted to macroscopic order and therefore coherently perform work. Interpreting the acoustic emission to represent crack initiation, propagation and sliding, and noting also that stress changes in elastic materials give rise to heat transfer (the thermoelastic effect),

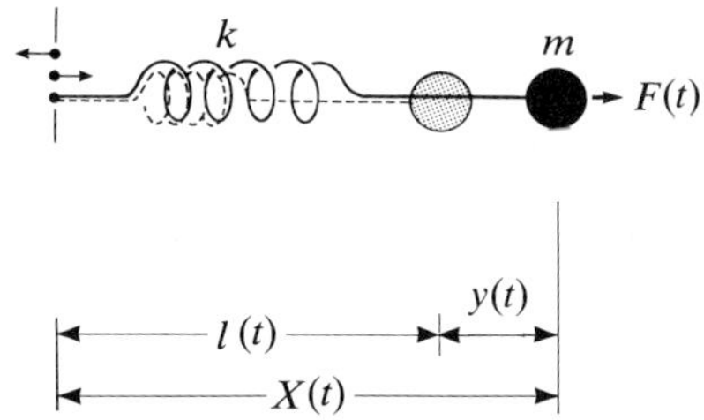

Figure 7. Single and mass system with variable equilibrium position $l(t)$. Modified after Dilts (1985).

the evolution of rock cores is seen to be consistent with the evolution of a dissipative dynamical system, and therefore also qualitatively consistent with the model of much larger lithospheric volumes developed in §3*b*.

Dilts (1985) represented processes such as microcracking by analogy to the elastic–plastic behaviour of a mass-spring system. This model can be used to illustrate the significance and evolution of the equilibrium state of the rock cores. Figure 7 shows Dilts's (1985) single mass-spring system, subject to an inhomogeneous force $F(t)$, in which the equilibrium position $l(t)$ can be varied independently of the motion of the mass m. The motion of the system may be described by

$$m\ddot{y} = -ky - m\,\mathrm{d}^2l/\mathrm{d}t^2 + F(t),$$

in which the term $m\mathrm{d}^2l/\mathrm{d}t^2$ represents the additional force generated by the changing equilibrium position. Now suppose that the equilibrium position is neither fixed nor arbitrarily variable but varies depending on the equations of motion, to represent yield of the spring when subject to either compressive or tensile forces. The motion can now be represented by

$$m\ddot{y} = -ky - m\,\mathrm{d}^2l(y,\dot{y})/\mathrm{d}t^2 + F(t).$$

Not only does the value of the equilibrium position $l(y,\dot{y})$ depend on the equations of motion, it in turn influences the equations of motion, so that the preceding expression represents a dynamical interplay between the two. Equilibrium becomes a dynamical quantity (state parameter). However, the actual value of the equilibrium position of mass m is irrelevant to the equations of motion because it merely introduces a constant force. The motion is influenced by the displacement from equilibrium y and the rate of change of this displacement, $\dot{y}$. The dynamical viewpoint illustrated by Dilts's analogy thus illustrates that the motion of the system is of overriding significance relative to the absolute value of the equilibrium state. This differs markedly from the more usual assumption that the state of a geological system is best described by quantifying the equilibrium state.

The approach to equilibrium involves fluctuation. Tan & King (1979) provided apparent experimental observations of such fluctuation. Dilts (1985) simulated fluctuation, including non-uniformity of the time-series, by calculations of the motion of multiple, connected mass spring elements. Multiple elements would better represent rock core, as does addition of a dissipative function (which was also incorporated by Dilts). It would also be appropriate to incorporate viscosity of the individual elements, but this is not a necessary condition for time-dependence. These additions would render the model more representative. However, the single mass-spring model provides a simpler basis to contrast the fundamental difference between descriptions which emphasize the equilibrium state of a geological system and those which emphasize the motion.

4. Conclusions

A number of different parameters and a range of forms of perturbation are used to determine rock stress state at a point location. Some aspects of the stress state determined by the different methods, notably the directions of the maximum and minimum horizontal stress, may be common to the different methods. However, based on the considerations given here, the stress magnitudes deduced from the different methods of measurement currently in use should not be assumed to be identical (even assuming precise, error-free measurements). The magnitude of the differences in a given situation may appear insignificant in engineering terms if treated simply as an error bar, or if members of individual determinations are averaged, thereby masking fluctuations. Nevertheless, it is desirable if not essential to develop a basis for reconciling the differences, or for selecting the measurement such that the measurement results most closely relate to the engineering or geological process in relation to which the stress measurements are conducted. The process giving rise to the quantitative value derived from a stress determination must be relatable to the relevant engineering or geological process. For example, a hydraulic fracture stress determination is ideal for designing pumping capacity and leak-off calculations for a stimulation treatment. However, it provides little or no information regarding the time-dependent closure of the fracture which can be inferred to result from evolution of the interaction forces in the walls of the stimulation fracture.

An understanding of the genesis of rock stress or some as yet unidentified generic relationship is necessary to the characterization of stress state from point determinations. Such determinations are often sparse and incomplete and consequently considerable extrapolation is required. Although any one of these factors may exert a dominant influence in a specific situation, observations suggest that stress is not simply related to lithology, structure or present-day conventional rock mechanical properties. Heterogeneity and gradients are characteristic of *in situ* stress state.

It is apparent that interaction forces should not be ignored. Field and laboratory results and theoretical considerations consistently suggest that interaction forces and surface forces are interdependent and not subject to superposition, which requires linearity. Interaction forces are characteristic of many rocks and probably of the entire accessible lithosphere. Internal strain changes and surface displacements are interdependent. It can be inferred that interaction forces influence the time-dependency of stress, strain and displacement and capacity for propagation of perturbations. Indeed, these considerations re-emphasize that rock stress is inextricably linked with rock constitutive properties. The two are interdependent expressions of the evolution of a nonlinear system. In practice they are separated, but the degree to which we can describe or predict the evolution of forces in geological systems may be seriously limited by this working approach.

For a number of reasons, but primarily because measurement involves perturbation which induces change, we may not be able to quantify internal force distributions. The conventional target of rock stress determination programmes, a quantification of surface forces or tractions which are assumed to be time-invariant, linearly related to internal forces and to represent the equilibrium state, is typically an approximation of unspecified inaccuracy. The considerations presented here, which accept the role of interaction forces, suggest that the equilibrium state is, strictly speaking, indeterminable from quasi-static analysis. Stress determinations may involve change to or from a reference state, typically the stress-relieved state. Yet the dynamics suggests that the new equilibrium state (stress-relieved overcore or hydrofracture on the point of closure) is a nonlinear function not only of the pre-test equilibrium state but also of the distance and history of distance from equilibrium during and after perturbation (by stress-relief or fracture opening/closing).

More attention to the dynamics of change rather than solely the assumed equilibrium state is justified for a number of reasons. First, the dynamics may be more relevant to both engineering design and our understanding of geological processes. The former is concerned with a response to perturbation, the latter a manifestation of system evolution. Understanding stress genesis in a nonlinear system is almost certainly necessary not only for extrapolation of individual stress determinations in time but almost certainly also in space. Second, recognition of the significance of the dynamics implies that new techniques of stress determination based on the dynamics of change may be feasible. Given the practical and economic constraints of existing techniques, any sound additions to the range would be welcome (and, as noted, may more closely relate to some of the engineering processes of interest and to geological processes). Third, the dynamics are presently or potentially observable in the field and the laboratory. Finally, the accessible lithosphere is a non-equilibrium nonlinear dissipative system. There is therefore the potential for self-organization and a considerable simplification of the apparent complexity.

The core experiments were conducted by a team from BP Research led by Dr J. T. Hagan. In addition to this valuable contribution, and the support of BP Research, we acknowledge stimulating discussions with Dr M. I. Asgian, Dr W. B. Bradley and Dr K. J. Heffer. The data in figure 1 were provided by R. J. Klein.

References

Anderson, R. N., Zoback, M. D., Hickman, S. H. & Healy, J. H. 1983 *In situ* stress and physical property measurements in the Cleveland Hills fault zone, Oroville, California. *Eos Trans. AGU* **64**, 834.

Arrowsmith, D. K. & Place, C. M. 1982 *Ordinary differential equations*. London and New York: Chapman & Hall.

Bak, P. & Tang, C. 1989 Earthquakes as a self-organised critical phenomenon. *J. geophys. Res.* B **94**, 15635–15637.

Bak, P., Tang, C. & Wiesenfeld, K. 1988 Self-organised criticality. *Phys. Rev.* A **38**, 364–373.

Barton, N. 1982 Hydraulic fracturing to estimate minimum stress and rock mass stability at a pumped hydro project. *Proc. Workshop XVII on hydraulic fracturing stress measurements, U.S. Geol. Surv. Open-File Rep.* 82-1075.

Bott, M. H. P. & Kusznir, N. J. 1984 The origin of tectonic stress in the lithosphere. *Tectonophys.* **105**, 1–13.

Chatwin, P. C. 1975 On the longitudinal dispersion of passive contaminant in oscillatory flows in tubes. *J. Fluid Mech.* **71**, 513–527.

Cornet, F. H. 1982 Analysis of injection tests for *in situ* stress determination. *Proc. Workshop XVII on hydraulic fracturing stress measurements, U.S. Geol. Surv. Open-File Rep.* 82-1075.

Cundall, P. A. 1990 Numerical modelling of jointed and faulted rock. *Proc. Int. Conf. on Mech. of jointed and faulted rock, Vienna, April.* (In the press.)

Dames & Moore 1978 *Geologic investigation, Nine Mile Point Nuclear Station Unit 2*, vol. 3. *Rock stresses.* Niagara Mohawk Power Corporation.

Dilts, G. A. 1985 The dynamical equilibrium: a mathematical model for the dynamical description of the permanent deformation of an elastic material caused by microfracture or plastic flow. Ph.D. thesis, University of Colorado.

Emery, C. L. 1963 Strain energy in rocks. In *State of stress in the Earth's crust. Proc. Int. Conf., June 13–14, Santa Monica, California.*

Grant, N. T. 1990 Episodic discrete and distributed deformation: consequences and controls in a thrust culmination from the Central Pyrenees. *J. struct. Geol.* **12**, 7, 835–850.

Haimson, B. C. & Lee, C. F. 1980 Hydrofracturing stress determinations at Darlington, Ontario. In *Underground Rock Eng: Proc. 13th Can. Symp. on Rock Mech.*, pp. 42–50.

Hallbjorn, L., Ingevald, K., Martna, J. & Strindell, L. 1990 A new automatic probe for measuring triaxial rock stresses in deep bore holes. *Tunnelling Underground Space Technol.* **5** (1/2), 141–145.

Harper, T. R. 1970 An investigation of stressed rock by observations of acoustic emission. M.Sc., Imperial College, University of London.

Harper, T. R. & Szymanski, J. S. 1981*a* Structural control of rock mass stress distribution. In *Mechanics of structured media. Proc. Int. Symp. on the Mechanical behaviour of structured media, Ottawa, May*, Pt. B, pp. 283–299. Elsevier.

Harper, T. R. & Szymanski, J. S. 1981*b* A perspective of geological characterizations required for subsurface nuclear waste disposal. In *Subsurface space* (ed. M. Bergman). *Proc. Int. Symp. Stockholm*, vol. 3, pp. 1345–1352. Pergamon.

Hast, N. 1967 The state of stresses in the upper part of the earth's crust. *Eng. Geol.* **2**, 1, 5–17.

Hickman, S. H. & Zoback, M. D. 1982 The interpretation of hydraulic fracturing pressure–time data for *in situ* stress determination. *Proc. Workshop XVII on hydraulic fracturing stress measurements, U.S. Geol. Surv. Open-File Rep.* 82-1075.

Hickman, S. H., Zoback, M. D. & Healy, J. H. 1988 Continuation of a deep borehole stress measurement profile near the San Andreas Fault. 1. Hydraulic fracturing stress measurements at Hi Vista, Mojave Desert, California. *J. geophys. Res.* **93**, 15183–15195.

Hiltscher, R., Martna, J. & Strindell, L. 1979 The measurement of triaxial rock stresses in deep boreholes and the use of rock stress measurements in the design and construction of rock openings. *Proc. 4th Int. Congr. Rock Mech., Montreux,* vol. 2, pp. 227–234.

Holzhausen, G. R. & Johnson, A. M. 1979 The concept of residual stress in rock. *Tectonophys.* **58**, 237–267.

Hooker, V. E. & Bickel, D. L. 1974 Overcoring equipment and techniques used in rock stress determination. *U.S. Bur. Mines Rep. Invest.* 8618, 32 pp.

Keilis-Borok, V. I. 1990 Introduction: non-linear systems in the problem of earthquake prediction. *Phys. Earth planet. Int.* **61**, 1–7.

Kry, P. R. & Gronseth, J. M. 1982 *In situ* stresses and hydraulic fracturing in the Deep Basin. Preprint, *33rd Ann. Tech. Mtg of Pet. Soc. of CIM, Calgary, June.*

Kurzweg, U. H. 1985 Enhanced heat conduction in fluids subjected to sinusoidal oscillations. *J. Heat Transfer* **107**, 459–463.

Leeman, E. R. 1969 The 'doorstopper' and triaxial rock stress measuring instruments developed by the CSIR. *J. S. Afr. Inst. Min. Metall.* **69**, 305–339.

Leeman, E. R. & Hayes, D. J. 1966 A technique for determining the complete state of stress in rock using a single borehole. In *Proc. 1st Cong. Int. Soc. Rock Mech., Lisbon,* vol. 2, pp. 17–24.

Lemos, J. V., Hart, R. D. & Cundall, P. A. 1985 A generalized distinct element program for modelling jointed rock mass. *Proc. Int. Symp. on fundamentals of rock joints, Bjorkliden, Centek, Lulea.*

Main, I. G., Peacock, S. & Meredith, P. G. 1990 Scattering attenuation and the fractal geometry of fracture systems. *Pageoph.* **133**, 283–304.

Martin, C. D. 1990 Characterising *in situ* stress domains at the AECL Underground Research Laboratory. *Can. Geotech. J.* **27**, 631–646.

McGarr, A. & Gay, N. C. 1978 State of stress in the Earth's crust. *A. Rev. Earth planet. Sci.* **6**, 405–436.

Nicolis, G. & Prigogine, I. 1989 *Exploring complexity.* New York: Freeman.

Nilson, R. H. & Lie, K. H. 1990 Double-porosity modelling of oscillatory gas motion and contaminant transport in a fractured porous medium. *Int. J. numer. analyt. Methods Geomech.* **14**, 565–585.

Prats, M. 1981 Effect of burial history on the subsurface horizontal stresses of formations having different material properties. *Soc. Petrol. Engng J.* 658–662.

Price, N. J. 1966 *Fault and joint development in brittle and semi-brittle rock.* Pergamon.

Ranalli, G. 1987 *Rheology of the Earth.* Allen & Unwin.

Schmitt, D. R., Bass, J. D. & Ahrens, T. J. 1985 Application of holographic interferometry to *in situ* stress measurement. *26th U.S. Symp. on Rock Mech., Rapid City, 26–28 June*, pp. 1055–1056.

Shamir, G., Zoback, M. D. & Cornet, F. H. 1990 Fracture-induced stress heterogeneity: Examples from the Cajon Pass scientific drillhole near the San Andreas Fault, California. In *Rock joints* (ed. N. Barton & O. Stephansson). Rotterdam: Balkema.

Smith, S. W. & Kind, R. 1972 Regional secular strain fields in Southern Nevada. *Tectonophys.* **14**, 57–69.

Sornette, D., Davy, P. & Sornette, A. 1990 Structuration of the lithosphere in plate tectonics as a self-organised critical phenomenon. *J. geophys. Res.* **95**, 17353–17361.

Strickland, F. G. & Ren, N.-K. 1980 Predicting the *in situ* stress state for deep wells using differential strain curve analysis. *Pet. Soc. CIM* Pap. no. 80-31-33.

Szymanski, J. S. 1989 Conceptual considerations of the Yucca Mountain groundwater system with special emphasis of the adequacy of this system to accommodate a high-level nuclear waste repository. U.S. Dept of Energy, Las Vegas.

Tan, T.-K. & Kang, W.-F. 1980 Locked in stresses, creep and dilatency of rocks, and constitutive equations. *Rock Mech.* **13**, 5–22.

Teufel, L. W. 1981 Strain relaxation method for predicting hydraulic fracture azimuth from orientated core. *Proc. SPE/DOE Low permeability gas reservoir symp.*, SPE/DOE 9836 Denver, 81–86.

Teufel, L. W. 1985 *In situ* stress state at the Mounds test well as determined by the anelastic strain recovery method. *SPE/DOE low permeability gas reservoirs symp.*, Denver, SPE/DOE 13896, 467–474.

Timoshenko, S. P. & Goodier, J. N. 1982 *Theory of elasticity.* McGraw-Hill.

Tullis, T. 1977 Reflections on measurement of residual stress in rock. *Pageoph.* **115**, 57–68.

Turcotte, D. L. 1989 Fractals in geology and geophysics. *Pageoph.* **121**, 171–196.

Velde, B., Dubois, J., Touchard, G. & Badri, A. 1990 Fractal analysis of fractures in rocks: the Cantor's dust method. *Tectonophys.* **179**, 345–352.

Voight, B. & St Pierre, B. H. P. 1974 Stress history and rock stress. *Proc. 3rd Congr. Int. Soc. Rock. Mech., Denver, A*, vol. 2, 580–582. Pergamon.

Wan, T., Tang, Y. & Zheng, W. 1989 Thermal structure of the lithosphere in Fujian, China. *J. geophys. Res.* **94**, 1888–1894.

Warpinski, N. R. 1986 Elastic and viscoelastic calculations of stresses in sedimentary basins. *Unconventional Gas Tech. Symp. of Soc. of Pet. Eng.*, Louisville, SPE 15243, 409–422.

Whitten, C. A., Green, R. & Meade, B. K. (eds) 1977 Recent crustal movements. *Tectonophys.* (Special Issue) **52**, 1–4.

Zheng, Z., Myer, L. R. & Cook, N. G. W. 1988 Borehole breakout and stress measurements. *Key questions in rock mechanics* (ed. P. A. Cundall, C. L. Starling & A. M. Starfield). *Proc. 29th U.S. Symp.* Rotterdam: Balkema.

Discussion

P. Hancock (*Bristol University, U.K.*). Accepting Dr Harper's observations as an account of what happens today, particularly in the context of variations in vertical stress of both magnitude and direction, why does he think that so many brittle geological structures in regions of simple tectonics in horizontal rocks are so uniformly orientated throughout such large areas? What does he understand by saying 'a long period of time'? Is this is a long period to a geologist or a long period to an experimentalist?

T. R. Harper. I am aware of Dr Hancock's observations not only of the uniformity of joint direction but also that this direction appears to correspond to the direction of maximum horizontal stress obtained by averaging a large number of *in situ* determinations. In this context we discussed the Syracuse–Oswego area in New York State, where a pronounced variation of stress magnitude and direction near the surface was recorded within a region of approximately 1 km^2 near Oswego. More than 100 successful determinations were made, and the variation was attributed to local structural evolution. I understand that when all these directional results were subsequently averaged, the result was a direction subparallel to the strike of joints in the region of Syracuse.

It may be relevant to recall the borehole breakout data we showed (figure 1). A substantial variation of the direction of horizontal stress with depth is inferred. When all the data points are represented as a rosette diagram (figure 1), they typically indicate a pronounced preferential direction (in some cases with approximately symmetrically disposed secondary directions). At larger scales, such as the U.K. North Sea region, some overall consistency, with local variation, is observed. From the non-equilibrium system viewpoint, which I have emphasized, fluctuation is to be expected not only in space but also in time. From this perspective, drawing such observations together, one can speculate that if the brittle geological structures developed progressively over a long period of time, then their finite resulting configuration may approximate to an averaged value of the stress directions which fluctuated in space and time throughout the period of growth of the structures.

I have made no attempt to quantify the period of time involved, but I used the adjective 'long' not in a geological context but to denote periods of typically 10^4–10^6 years. In high-energy flux areas, the lower bound would be reduced by an order of magnitude or more. Please note, however, that these numbers are only a generalization, not only because they are not rigorously derived, but also because they have little meaning without specifying accompanying magnitudes. Fluctuation of geological systems can be observed at very much shorter timescales (less than 10 years), such as by geodetic relevelling. The motion of geological systems appears to occur at a wide range of scales, and one is tempted to speculate that the spatiotemporal fluctuations may be characteristically self-similar.

A. Batchelor (*Geoscience Ltd, U.K.*). Dr Harper made the general conclusion about the vertical stress not being a function of integrated density and showed an example from Canada. Does he have any more evidence for that? My experience of measuring stresses at depth is that once you get below 400 or 500 m in an area of not severe topography the vertical stress can indeed be approximated by the density. In the North Sea, in a block not so far from the one that he mentioned, I would argue that we saw evidence that vertical stress could be estimated from integrated density.

T. R. Harper. Although for many engineering purposes it is often adequate to approximate the vertical stress by depth-integration based on density logs, I have attempted to emphasize that it is not always helpful to assume that the vertical stress can be accurately represented by such an assumption. I have in mind two particular situations in which this assumption would be unhelpful. The first is high-energy transfer rate environments, in which the stress state is evolving relatively rapidly and one is concerned with quantifying the implications, such as any fluid flow driven by evolving deformation gradients. This situation can be inferred to pertain at Yucca mountain, Nevada, a candidate nuclear waste repository site with active faults and high heat-flow gradients.

The second situation is that in which the objective is to understand lithospheric processes. Perhaps the presence of gradients of vertical stress departing from that calculated from overburden weight is easiest to appreciate by reference to the structural evolution of an evolving high-angle normal fault, for which an evolving gradient of fault-parallel normal stress may be readily inferred on genetic grounds.

Observations suggest a spatial variability of stress which, when averaged, suggests some consistency. Regrettably, determinations of vertical stress are far less numerous than of horizontal stresses and so (bearing in mind also the problems of measurement error) it is not an easy matter to test the proposed variability in relation to vertical stress. Nevertheless, Dr Batchelor will recall that the overcoring measurements with which he was involved at South Crofty Mine showed precisely this effect: I believe six CSIRO measurements at a depth of about 800 m yielded values of the near-vertical intermediate principal stress which varied substantially (*ca.* 9 MPa), yet the average of the six values, which was aligned vertically, yielded, as you have implied, an acceptable approximation of the magnitude which would be obtained from the integrated density approximation.

Hiltscher *et al.* (1979) also observed a variability of vertical stress. Other references citing a departure from the calculated lithostatic value are given in the written paper. So, to summarize, it appears that integration of density logs is adequate for many but not all engineering purposes, but that there is sufficient evidence to warn us that it is neither accurate, helpful nor supportable on genetic grounds to assume that the vertical stress is always equal to the integrated weight of overburden.

S. Murrell (*University College, London, U.K.*). In the thermal cycling experiment in the laboratory, was allowance made for the slow diffusion of heat in rock samples where one would expect there to be substantial strain lag effects? Could this be the possible explanation for the difference between steel and rock?

T. R. Harper. The 100 mm diameter core and control samples were maintained at the elevated and reference temperatures for sufficient time for temperatures to become uniform, typically not less than one day. Obviously transient temperature and stress gradients are imposed, but we confirmed that the cores were uniformly heated for the majority of each temperature cycle by continuously recording the output from thermocouples with sensors located in the centre of the samples.

N. Kusznir (*University of Liverpool, U.K.*). One of the issues concerning many of the people here is to what extent we can use local measurements of stress to actually get a feel for the horizontal force distribution in the lithosphere on a global scale. What is the implication of the variability, both temporally and spatially, that is observed for our mapping of what is often called 'global stress', but what is probably an approximation to global lithosphere force?

T. R. Harper. In addition to the variability of the results of stress determinations that I have discussed, geological, geomorphological and geodetic observations demonstrate that spatiotemporal fluctuation extends across a wide range of scales. These data are consistent with the concept of the lithosphere as a non-equilibrium thermodynamic system, and suggest that this variability should apply throughout the lithosphere. Therefore, we should be cautious of extrapolations based on the assumption of steady-state conditions.

Near-surface stress determinations are potentially a useful guide to horizontal force distribution in the lithosphere only if first we make allowance for the spatial variability in the stress direction determinations and, second, develop a genetic model upon which to base the extrapolation to depth. Also, the concept of stress may be inappropriate when gradients are characteristic of a wide range of scales. This becomes readily apparent at the microscopic level, at least in clastic and crystalline rocks, at which scale the use of force is more helpful. It is also particularly apparent at the other extreme of spatial scale, involving large volumes of lithosphere, largely inaccessible, at which scale the lack of resolution potentially achievable also suggests that the use of force may be more suitable than stress.

A stress measurement profile to mid-crustal depth in KTB scientific drilling project, southeastern Germany: scientific questions and technological challenges

By Karl Fuchs[1], Mark D. Zoback[2]† and Fritz Rummel[3]

[1] *Institut für Geophysik, Karlsruhe Universität, 7500 Karlsruhe 21, F.R.G.*
[2] *Department of Geophysics, Stanford University, Stanford, California 94305, U.S.A.*
[3] *Institut für Geophysik, Ruhr-Universität, 4630 Bochum, F.R.G.*

Although many estimates of crustal stress levels at seismogenic depth have been made on the basis of laboratory data and simple faulting theory, the *Kontinentales Tiefbohr* (KTB) deep drilling project offers the first opportunity to directly measure at mid-crustal depth the absolute magnitudes of crustal stresses, the uniformity of stress changes with depth, the influence of pore pressure on stress magnitude and the magnitude and orientation of tectonic stresses in the vicinity of the brittle–ductile transition. The planned 10–12 km depth of the KTB *Hauptbohrung* (main borehole) will penetrate those depths where the shear stresses in the crust are highest and where the greatest discrepancy will exist between observations and predictions if either 'Byerlee's law' is not applicable to the mid-crust or near-lithostatic pore pressure exists which substantially weakens the crust.

Consistent with measurements in the upper 3 km at a number of sites around the world, stress magnitudes measured to 3 km depth in the 4 km deep KTB *Vorbohrung* (pilot hole) were found to be in agreement with Byerlee's law. However, there are a number of problems with simply extrapolating such data to mid-crustal depth. First there is now a substantial body of information that indicates that plate boundaries move at shear stress levels appreciably lower than those predicted by Byerlee's law. As we do not yet understand the origin of this weakness, it is not clear whether the processes responsible for it might also apply to plate interiors. Second, in a number of intraplate continental areas, it is now clear that the state of stress is closely related to plate-driving processes. However, the magnitude of plate-driving forces is considerably less than that required to cause intraplate deformation, especially in regions of reverse and strike–slip faulting like eastern North America. Third, it is essentially impossible to predict the relationship between pore pressure and stress magnitude at the depths where the brittle–ductile transition is approached. The onset of creep and ductile deformation would seem to be associated with the closure and healing of microcracks resulting in pore fluids trapped in isolated cracks and pores. This would cause a breakdown of the effective stress principle and a marked increase of crustal strength (making the discrepancy between the magnitude of plate-driving forces and crustal strength even more severe). However, geologic evidence from veining and the nature of earthquake focal plane mechanisms from some earthquake swarms, indicates that in some places, at some times, near lithostatic

† Current address: Geophysical Institute, University of Karlsruhe, 7500 Karlsruhe 21, F.R.G.

Phil. Trans. R. Soc. Lond. A (1991) **337**, 25–28
Printed in Great Britain

fluid pressures exist at mid-crustal depth. For these reasons we believe that continuous determination of stress magnitudes and orientation to mid-crustal depth are key experiments of the KTB project and of broad-scale scientific interest.

Techniques for measurement of *in situ* stress magnitudes standardly used at shallower depth are difficult to impossible to apply at mid-crustal depth. For this reason, we have developed an integrated stress measurement strategy that involves (1) a modification of the conventional hydraulic fracturing technique at relatively shallow (*ca.* 4–6 km) depth utilizing aluminium packers in small-diameter pilot holes drilled periodically at the bottom of the *Hauptbohrung*, (2) hydraulic fracturing through cemented casing that will potentially make measurement of the magnitude of the least principal stress possible to the total depth of the hole, and (3) estimating the magnitude of the greatest horizontal principal stress from the detailed analysis of compressional and extensional wellbore failures. This requires utilization of knowledge of the magnitude of the least principal stress (made available from the hydraulic fracturing tests), rock strength and pore pressure.

Discussion

P. England (*Oxford University, U.K.*). First it seems to me that Professor Zoback's difficulty with Byerlee's law comes about through estimates of stress which he makes in the upper mantle and those estimates depend more strongly on temperature at the Moho than anything else, which I think he must admit is rather indeterminate. So possibly the problem really resides in the estimates of stress from Byerlee's law in the middle crust. When he said that pore pressure in the middle crust is essentially equal to static pressure this would imply a great variety of focal mechanisms. Does that implication still hold up if there is a kinematic constraint on the deformation, if it is imagined that the whole of the crust is being driven by distributed tractions applied to its base by a stronger substrate? Would he not then get consistent sets of focal mechanisms and that would then let him postulate high pore pressures in the crust?

M. D. Zoback. I agree that the temperature in the upper mantle is the predominant parameter controlling mantle strength, and thus how much plate-driving force the mantle is 'carrying'. However, the problem I alluded to, the discrepancy between the overall strength of the lithosphere and the amount of force available to cause lithospheric deformation, is primarily the result of applying Byerlee's law with hydrostatic pore pressure through the brittle part of the upper crust in compressional environments. Thrust faulting to a mid-crustal depth of about 15 km using Byerlee's law with hydrostatic pore pressure requires a cumulative force of about 7×10^{12} N m^{-1} which is about 50% greater than the forces generally thought to be available to deform the lithosphere. This is independent of temperature as it involves no force being carried in the ductile lower crust or upper mantle. So indeed, the question is the validity of the assumption that Byerlee's law is applicable with hydrostatic pore pressures. The way out of this problem is to make the brittle crust much weaker. An extremely low coefficient of friction (incompatible with Byerlee's law) or a very high (near lithostatic) pore pressure are two ways to do this. The question of whether the uniformity of the *P*- and *T*-axes in intraplate areas is the result only of kinematic constraints is an interesting one. For this to be true it implies that earthquake focal-plane mechanisms tell us only about strain (and nothing about stress). This can be true only if the frictional strength of the faults is close to zero. So, basically, we reach the same conclusion no matter how we interpret the deep focal-plane mechanisms. If they are stress indicators the frictional strength of the crust appears to be too high; if they are only strain indicators the frictional strength of the crust must be close to zero.

N. KUSZNIR (*Liverpool University, U.K.*). The strain rates shown are exceedingly rapid, 10^{-15} is 100% stretching in 30 Ma. It is a respectable rate for forming an extensional basin. By the time 10^{-11} is reached I don't think any process on the Earth goes that fast. Obviously if one reduces the strain rates, one reduces the force quite substantially.

M. D. ZOBACK. The examples of strength envelopes I was using were made for plate boundaries and not for intraplate areas and thus the strain rates were indeed too high for an intraplate region. However, as in my response to the question by Dr England, the assumed strain rates (like temperature) only affect the strength of the ductile lower crust and upper mantle and the problem of lithospheric strength against lithospheric force comes from what is happening in the brittle part of the upper crust, from the applicability of Byerlee's law and from hydrostatic pore pressure.

M. H. P. BOTT (*Durham University, U.K.*). The heat flow Professor Zoback presented is about the continental-shield heat flow and the gross strain rate is certainly very small in such areas. The same of course applies to oceanic lithosphere. Is this the reason why these regions are not deforming and why they have strength and the movements are concentrated in the hot regions? Further would this suggest in field areas that the upper part may be in frictional equilibrium, while in the lower part the stresses are much less?

M. D. ZOBACK. I think this suggestion is quite likely. In fact, other calculations I have done suggest that it may be that only in areas of very low heat flow (like shields and old ocean basins) that the upper mantle can carry much plate-driving force (and thus the crust does not have to!). So indeed, perhaps this is why deformation rates are so low in such areas.

T. HARPER (*BP Research Centre, Sunbury, U.K.*). I think we know enough about the evolution of faults to appreciate, from fairly recent studies over the past five or ten years, that the state of stress evolves around any particular fault which has a finite geometry and a finite extent and, as fluid flow is associated with it, it is a quite complicated evolution. It seems to me that if Professor Zoback's measurements are consistent with a long structure with one block sliding over the other, either he is making his measurements remote from the sliding fracture, or the idea of an evolving stress distribution round a finite fault somehow does not apply in his case. He had a scatter of something like 60° over quite a large amount of the data. I think that we need to try to define what is a significant variation. Can he attach a number to this term?

M. D. ZOBACK. The interesting thing about stress measurements near major active faults is both their uniformity and variability. For example, near the San Andreas fault in central California, the direction of maximum horizontal stress from scores of wellbore breakouts and earthquake focal-plane mechanisms adjacent to the fault, indicate a remarkably uniform direction of maximum horizontal stress oriented nearly perpendicular to the strike of the fault. At the same time, detailed studies of stress orientation to 3.5 km depth in the Cajon Pass research borehole adjacent to the San Andreas fault in southern California show a well-established mean stress orientation (approximately perpendicular to the local strike of the San Andreas fault) with a standard deviation of 19°, but numerous small-scale fluctuations due to motion on small faults penetrated by the borehole. The only places where large scatters of 60° exist (such as those Dr Harper refers to), are associated with these extremely localized stress perturbations. The average stress field is fairly uniform.

Determining contemporary stress directions from neotectonic joint systems

By P. L. Hancock

Department of Geology, University of Bristol, Wills Memorial Building, Queen's Road, Bristol BS8 1RJ, U.K.

A neotectonic joint is a crack which propagated in a tectonic stress field that has persisted with little or no change of orientation until the present day. Investigating neotectonic joints is of value because the approximate orientation of the contemporary stress field can be inferred from them.

Although exposed neotectonic joints in the flat-lying sedimentary rocks of some cratons are related to regional stress fields, their initiation and propagation occurred close to the Earth's surface. For example, neotectonic joints in the centre of the Ebro basin (N. Spain) preferentially developed in a thin, near-surface channel sited within a sequence of weak Miocene limestones underlying the upper levels of plateaux. The Ebro basin joints strike uniformly NNW–SSE throughout an area of at least 10000 km^2 and they are parallel or subparallel to the direction of greatest horizontal stress extrapolated from *in situ* stress measurements and fault-plane solutions of earthquakes.

1. Preamble

This paper has three aims. Firstly, to review the general attributes of exposed neotectonic joints, secondly, to demonstrate that in S. England/N. France and the Ebro basin (Spain) the directions of greatest horizontal principal stress that were inferred from neotectonic joints are parallel to those determined later from geophysical observations, and thirdly, to discuss why neotectonic joints in the Ebro basin are better developed in weak limestones cropping out in the upper parts of plateaux.

In this paper the word neotectonic is used to indicate that a fracture was propagated in a tectonic stress field that has persisted with little or no change of orientation until the present day (Hancock & Engelder 1989). A fracture is called a joint if, at the scale of observation possible in the field, it is barren and there is no measurable offset related to shear, dilation or pressure solution. Because joints are the most abundant of non-penetrative geological structures they are of great potential value for tracking the orientations of principal stress axes at the time of failure. Confidence in the value of joints as stress indicators is greatest where they are uniformly arranged throughout a large area of flat-lying rocks.

2. Fracture classes and joints

On the basis of theory, rock mechanics and field observations, many geologists recognize three classes of fractures; extension fractures, hybrid-shear fractures and

Phil. Trans. R. Soc. Lond. A (1991) **337**, 29–40

Printed in Great Britain

Figure 1

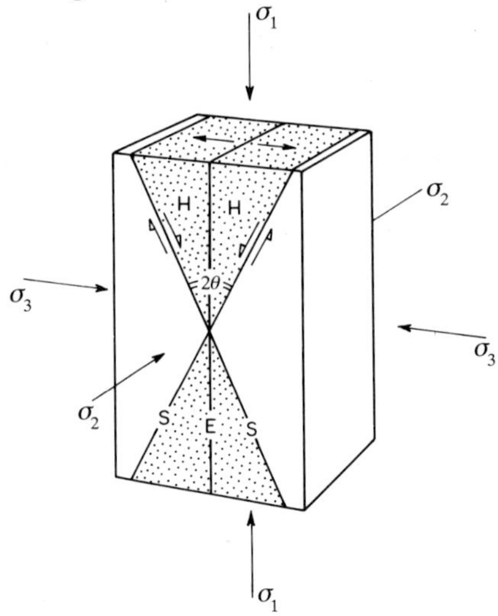

Figure 2

Figure 1. Relationships between fracture class and principal stress axes during the failure of brittle intact rocks. E, single extension fracture; S, conjugate Coulomb-shear fractures; H and stipple, field within which conjugate hybrid-shear fractures form at small angles to σ_1. Principal stresses are $\sigma_1 > \sigma_2 > \sigma_3$. 2θ is the conjugate shear angle.

Figure 2. A plumose marking with a horizontal axis on a NNW-striking neotectonic extension joint cutting a Miocene chalky limestone, Sancho Abarca, Zaragoza Province, Ebro basin, Spain. The joint, which is about 1 m high, propagated from right to left. View to the west.

Coulomb-shear fractures (Price & Cosgrove 1990). Figure 1 illustrates relationships between fracture classes and the orientations of principal stress axes, assuming the rock is brittle and does not contain pre-existing fractures. All the joints to be discussed in this paper formed in brittle rocks and many of them propagated in intact rocks. Even those that developed in already jointed rock possess many characteristics in common with those that formed in intact rocks.

Many joints are interpreted as extension fractures. From such joints the orientation of the σ_3 axis can be determined knowing that it was perpendicular to the joints. Unless a joint displays a plume axis it is not possible to determine the orientations of σ_1 and σ_2 within the plane. Plume axes and individual plume components (figure 2) develop parallel to σ_1 within a joint plane (Kulander *et al.* 1979).

Whether some joints are hybrid-shear or Coulomb-shear fractures is a controversial issue (see Pollard & Aydin (1988) and Price & Cosgrove (1990) for conflicting views). In the view of Price & Cosgrove (1990) and Hancock (1985) there is abundant field evidence in favour of the formation of conjugate hybrid-shear joints. From conjugate sets of joints it is possible to infer the orientations of all three principal stresses knowing that the acute and obtuse bisectors between sets yield σ_1 and σ_3, respectively. Although many joints belong to well-defined sets within which the angular dispersion of planes is small, others define a coaxial angular continuum enclosing a maximum 2θ angle of about 45°. Hancock (1986) has interpreted such continua as comprising a spectrum of extension and hybrid-shear fractures.

From the perspective of comparing the orientations of principal stress axes inferred from joint sets with those known from *in situ* stress measurements it is usually possible to compare only directions of horizontal stress. In this account, the direction of greatest horizontal stress is called S_H, while that of the least horizontal stress is referred to as S_h. S_h and σ_3 are identical in the examples discussed here, but S_H is equivalent to σ_1 in some situations and σ_2 in others.

Plumose markings and arrest lines are not only valuable indicators of stress axis orientations but they also allow the propagation sequence of the joint bearing them to be established. For example, only rarely do joint surfaces display more than about six arrest lines. This means that the final dimensions of many joints are achieved in no more than about seven propagation events during which significant rotations of the axes of a regional stress field are unlikely to occur.

3. Common attributes of exposed neotectonic joints

Exposed joints that have been interpreted as neotectonic have been described from Devonian rocks in the Appalachian Plateau of New York State, Ordovician rocks in the Valley and Ridge Province of Pennsylvania, Late Cretaceous–Early Tertiary rocks in southern England and northern France, Neogene rocks in eastern Arabia, and Miocene rocks in the Ebro basin of northern Spain (Engelder 1982, 1985; Bevan & Hancock 1986; Hancock 1987; Hancock & Engelder 1989). In addition, Holst & Foote (1981) have described joints cutting Devonian rocks in the Michigan basin that strike parallel to the contemporary direction of S_H. On the basis of these studies the following inventory of fundamental and field attributes of *exposed* neotectonic joints has been compiled.

1. Neotectonic joints either strike parallel to S_H or symmetrically enclose a small angle about S_H where its orientation is known from geophysical data. Parallelism or subparallelism of a joint set with S_H does not, by itself, indicate that a joint is neotectonic.

2. Most exposed neotectonic joints are thought to have propagated in the uppermost 500 m of the brittle crust because they are absent in cores taken from greater depths (Engelder 1985).

3. Regionally extensive neotectonic joint sets within a tectonic domain of several thousand square kilometres generally strike in a uniform direction.

4. Neotectonic joint systems are geometrically simple, mainly consisting of a single set of vertical systematic joints (figure 3*a*). Single sets are locally replaced by either conjugate sets at small dihedral angles to each other or a joint spectrum. Single sets of systematic neotectonic joints generally comprise extension fractures that were propagated when S_h was perpendicular to a set and S_H was parallel to its strike. When steeply inclined joints in conjugate sets or spectra were formed, S_h was slightly oblique to joint planes while S_H remained parallel to their strike (figure 3*b*). During the formation of vertical joints in conjugate sets or spectra, S_h and S_H subtended large and small angles, respectively, with the joint planes (figure 3*c*). The dominance of either extension or hybrid-shear joints enclosing small dihedral angles indicates that differential stresses were small during jointing.

5. Neotectonic joints are the youngest systematic joints (Hodgson 1961) in a rock although individual neotectonic joints may be linked by small irregular non-systematic joints abutting them (figure 3). Where neotectonic joints have formed in rocks containing older and sealed systematic joints they either cut or abut them. Regionally extensive and exposed neotectonic joints do not define orthogonal networks of roughly coeval systematic sets. Such networks are common where

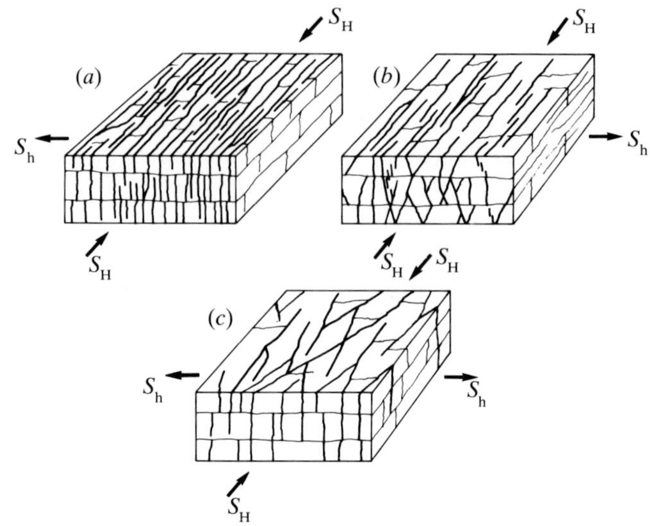

Figure 3. Characteristic neotectonic joint systems. (*a*) Single set of systematic vertical extension joints (heavy lines) linked by non-systematic cross-joints (thin lines). (*b*) Spectrum of systematic vertical extension joints and steep hybrid-shear joints linked by non-systematic cross-joints. (*c*) Spectrum of systematic vertical extension and hybrid-shear joints linked by non-systematic cross-joints. S_H, greatest horizontal stress; S_h, least horizontal stress (after Hancock & Engelder 1989, fig. 5).

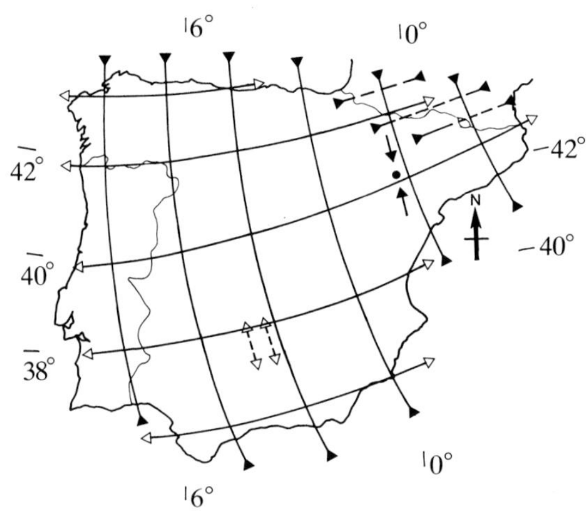

Figure 4. Contemporary regional stress trajectories (continuous and pecked lines) in the Iberian peninsula extrapolated from *in situ* stress measurements at nine sites, earthquake focal mechanisms, at four sites and the analysis of recent faults cropping out at four sites or visible on *Landsat* images at three sites (after Gonzalez de Vallejo *et al.* 1988, fig. 4). The direction of the greatest horizontal stress axis inferred by Hancock (1987, fig. 11) from the strike of neotectonic joints in the Candasnos area of the Ebro basin is also shown by arrows external to a solid circle. Projection: conical with two standard parallels.

ancient (i.e. palaeotectonic) joints have been exposed as a result of uplift and exhumation (Hancock *et al.* 1987). However, local networks of coeval orthogonal joints are formed adjacent to some neotectonic normal faults (Stewart & Hancock 1990). The general geometry of neotectonic joint systems is identical in both intact and previously fractured rocks although the orientation of a neotectonic joint may be locally deflected within a few centimetres of an old one.

6. Veins parallel to and of the same age as neotectonic joints are rare. Palaeotectonic joints are commonly accompanied by parallel vein sets. The absence of veins parallel to exposed neotectonic joints suggests that the joints formed when the fluid pressure to confining pressure ratio was low and significantly less than one.

7. Neotectonic joints commonly cut several beds. The principal exception to them being multilayer fractures is within sequences containing weak layers, such as clays or marls; joints in these sequences do not pass through the weak layers. Neotectonic joint zones containing several closely spaced joints within a narrow zone (centimetres wide) are common in rocks containing older joints. Large multilayer neotectonic joints are usually slightly gaping.

4. Successful estimates of the direction of S_H from neotectonic joints

In 1986 Bevan & Hancock reported the presence of NW-striking joints and small normal faults cutting Late Cretaceous chalks and Palaeogene sands and clays in S. England and N. France. They attributed the formation of the fractures to the influence of a stress field within which the regional direction of horizontal extension was orientated NE–SW, and they noted that the strike of the fractures was parallel to the direction of S_H determined from *in situ* stress measurements and fault plane solutions of earthquakes external to their study-area. One year after the appearance of Bevan & Hancock's article, Brereton & Evans (1987) published the results of an analysis of borehole breakout data that had become available to the British Geological Survey. Many of the boreholes are sited within the area surveyed by Bevan & Hancock (1986): from them Brereton & Evans (1987) concluded that the direction of S_h was mainly NE–SW (see Hancock & Engelder 1989, fig. 2, for details).

NNW-striking joints cutting flat-lying Miocene rocks near Candasnos in the east-central part of the Ebro basin (N. Spain) were interpreted by Hancock (1987, fig. 4) as neotectonic using the field criteria listed in §3. On the basis of *in situ* stress measurements by overcoring, earthquake focal mechanisms, and analyses of recent faults visible at outcrop or on *Landsat* images, Gonzalez de Vallejo *et al.* (1988) published a map of the Iberian peninsula showing smoothed contemporary stress trajectories. They showed the direction of S_H in the Candasnos area within 5° of the average strike of the joints that were first reported by Hancock (figure 4).

5. Neotectonic joints in the Ebro basin

Figure 5 shows average strikes of joints surveyed by the author in the flat-lying Miocene rocks of the Ebro basin, north Spain. The only fractures disturbing these horizontal rocks are small faults and joints. Most of the faults are normal, and those in a graben-field near Tudela strike NNE–SSW and were formed between the Aquitanian and Tortonian stages according to Garcia Prieta & Simón-Gómez (1986). The joint sets depicted in figure 5 are mainly vertical or nearly vertical but they are not all of the same age or type. Those striking WNW–ESE along the northern fringe of the basin are extension joints that Turner & Hancock (1990) interpret as being related to flexural loading of the basin during the early Miocene. NNE–SSW striking joint in the same tract are extensional cross-joints related to the WNW-striking Pyrenean joints.

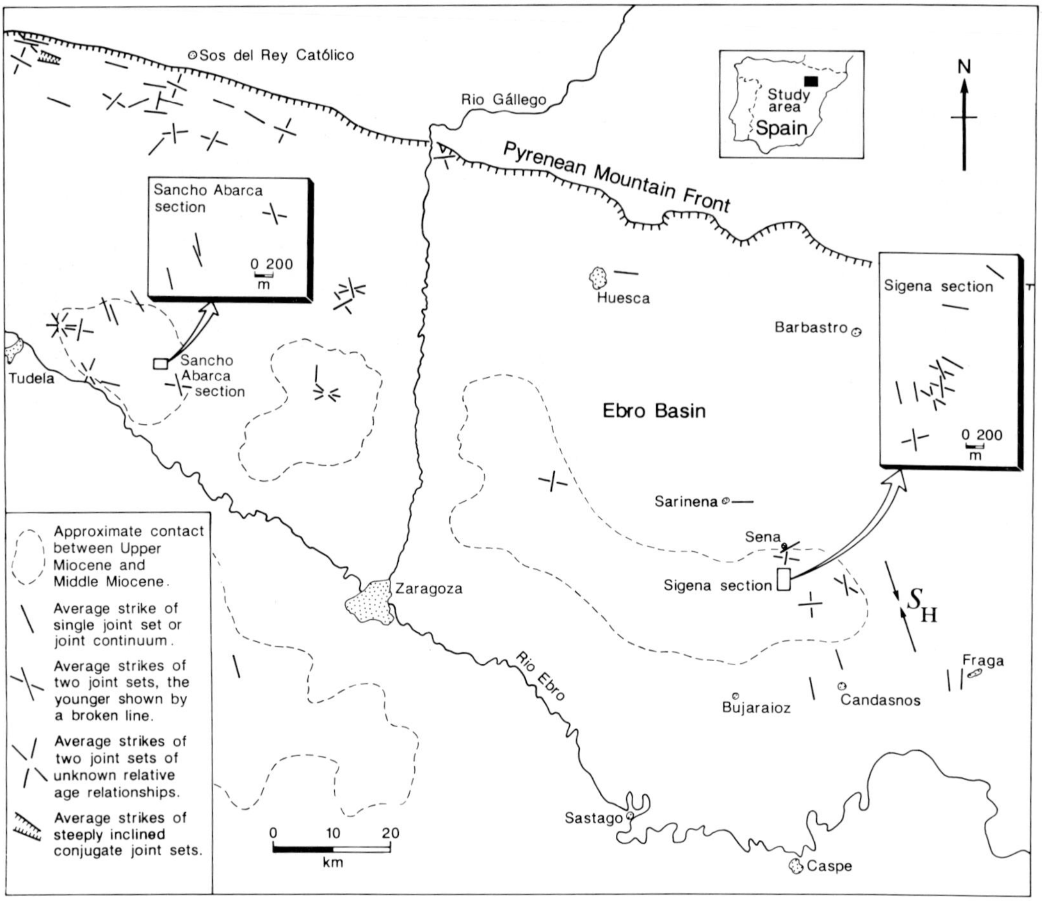

Figure 5. Average strikes and relative age relationships of joint sets in part of the Ebro basin, N. Spain. S_H, contemporary direction of the greatest horizontal stress according to Gonzalez de Vallejo *et al.* (1988, fig. 4). Upper/Middle Miocene contact after Riba, Puigdefabregas and Quirantes (in Julivert 1978, fig. 4.8). Strikes of joints adjacent to the Pyrenean mountain front derived from data in Turner & Hancock (1990, figs 4–6).

In the centre of the basin, joints of another trend cut lacustrine and fluvial rocks that Riba *et al.* (in Julivert 1978) dated as middle to late Miocene. Because these rocks are younger than the latest Pyrenean events, the joints that cut them must be related to younger deformation phases. Of particular interest in the context of this paper are NNW-striking joints cutting the youngest Miocene rocks cropping out in plateaux and mesas. Because NNW-striking joints abut NNE-striking normal faults and parallel shear fractures (figure 6) it is concluded that the joints are post-Tortonian (i.e. late Miocene) structures. Furthermore, because they are of late Cenozoic age and strike parallel or subparallel to the direction of S_H it is concluded that they are neotectonic. Nine attributes of these joints are noteworthy.

1. As reported in §4, the direction of S_H inferred from the joints is within 5° of that extrapolated from independent geophysical evidence.

2. The strike of joints, or the direction of the acute bisector between vertical conjugate hybrid-shear joints, is generally within 15° of an average strike of 165° throughout an area of about 10000 km^2 (figure 5).

Figure 6

Figure 7

Figure 6. Vertical NNW-striking neotectonic extension joints abutting moderately inclined, NNE-striking shear fractures (sloping down to the right) cutting Miocene chalky limestones at Sancho Abarca, Zaragoza Province, Ebro basin (N. Spain). The joints are younger than the late Miocene shear fractures. The exposure is about 2 m high. View to the north.

Figure 7. Closely spaced and planar NNW-striking neotectonic joints cutting Miocene chalky limestones at Sancho Abarca, Zaragoza Province, Ebro basin (N. Spain). Individual joint planes are picked out by the dark shadows. View to the northeast.

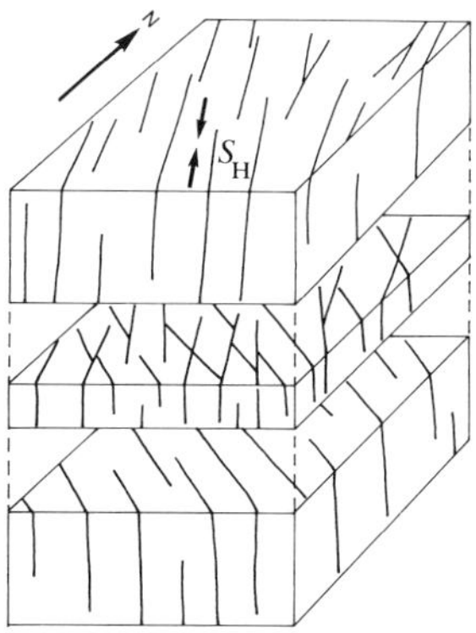

Figure 8. Exploded block diagram illustrating relationships between dominant systematic joint sets in the neighbourhood of the 20 m thick 'joint transition zone' within the Sigena section (see figure 5) in the Ebro basin. The highest beds (shown as one) are cut only by NNW-striking neotectonic joints. In the transition zone there are both NW- and NNW-striking joints. Layers immediately beneath the transition zone contain only NW-striking joints. S_H, direction of contemporary greatest horizontal stress according to Gonzalez de Vallejo *et al.* (1988, fig. 4).

3. A set of vertical extension joints is dominant at most localities but locally it is replaced by either steeply inclined or vertical conjugate hybrid-shear joints in sets or spectra.

4. NNW-striking joints are the youngest systematic joints in the study area.

5. The morphology of NNW-striking joints in limestones is the same throughout the area, thus enabling them to be identified by a field criterion distinct from azimuth. Although individual parts of joints are smooth many surfaces are composite and stepped. Many joints that strike in directions other than NNW–SSE are irregular.

6. A few joint planes in homogeneous limestones display plumose markings with horizontal plume axes (figure 2).

7. Systematic NNW-striking joints are absent in the weak clays and marls interbedded with chalky limestones.

8. Veins parallel to NNW-striking joints are absent or rare.

9. NNW-striking joints are best developed (i.e. most closely spaced and planar) within a sequence of chalks that underlie mesas capping plateaux (figure 7). NNW-striking joints become less closely spaced and less planar towards the tops and bases of most mesas. Beneath the higher levels of all mesas and in the youngest rocks, NNW-striking joints are the only systematic joints. These relationships are especially well displayed in the Sigena section (figure 5), which starts at about 250 m above sea level and rises to nearly 600 m. Joints striking NNW–SSE suddenly appear at the 490 m level and are present in all limestones up to the top of the escarpment. For about 20 m above their first appearance they co-exist with NW-striking joints (figure 8). These joints are identical in strike to those cutting beds immediately below and lacking NNW-striking joints. Thus in the Sigena section there is a thin 'joint transition zone' containing joints of both strikes, but otherwise joints of either NW–SE or NNW–SSE strike are restricted to the lower or upper parts of the section, respectively.

Three aspects of the exposed neotectonic joints in the Ebro basin require comment from the perspective of understanding their origin.

1. Why are the joints uniformly orientated throughout an area of 10000 km^2 and within a sequence containing limestones that are separated from each other by unjointed mudstones? Although the limestones containing the joints are of uniform thickness on the scale of an outcrop they are lenticular on the scale of a few hundred metres and were, thus, never, continuous across the basin. Hence it is impossible that they acted as basin-wide stress guides. It is more probable that the stresses responsible for generating the joints in the limestones reflect small strains in the enclosing muddy rocks. In their turn, these strains probably mirror small regional strains of uniform orientation beneath the centre of the basin, which is distant from the boundary conditions imposed from the framing mountain ranges (Simón-Gómez 1989).

2. Why, in the centre of the basin, do the exposed neotectonic joints, and hence the inferred direction of S_H (locally σ_1), trend NNW–SSE? Because the direction of S_H is oblique to the trend of both the Iberic and Pyrenean mountains it is unlikely to reflect the dominant direction of shortening in either of those domains. A more likely explanation, judging from figure 4, is that it reflects a large regional stress field affecting all of Iberia.

3. Why are neotectonic joints in the Ebro basin best developed in a relatively thin near-surface channel beneath the summits of plateaux and mesas? Two explanations can be put forward. One is that neotectonic joint development is controlled by lithology. In favour of this hypothesis is the observation that the most closely spaced and planar joints are restricted to fine-grained homogeneous marly chalks with a uniaxial strength of about 20 MPa (as estimated from point load tests by using the method of Bock & Franklin (1972)). However, the limitation of neotectonic joints to chalky limestones does not explain why they die out upwards in chalks which at the top of the sequence are little different to those 100 m below.

An alternative explanation for the restriction of exposed neotectonic joints in the Ebro basin to a near-surface channel is that for their development both a thin cover was required to provide a confining pressure, and they had to be uplifted to a level within which, after exhumation, tensile or hybrid-shear failure could occur without the aid of abnormally high fluid pressures. Thus neotectonic joints are absent at the tops of some mesas because there was either insufficient cover or there has been little exhumation. Neotectonic joints are, however, exposed at the tops of more denuded mesas from which a former cover has been removed. The observation that neotectonic joints are absent or less well developed in the lower slopes of some mesas is probably related to the rocks of such settings not having been sufficiently extended. Uplift is likely to have been the principal mechanism leading to horizontal extension. It is otherwise difficult to understand why NNW-striking neotectonic joints have not been superimposed on older joints cutting rocks lower in the sequence (e.g. see figure 8). In the context of Engelder's (1985) genetic classification of joints, the exposed neotectonic joints in the central part of the Ebro basin are typical *unloading joints* formed after burial, uplift and exhumation.

The work of Narr & Burress (1984) on the formation of fractures observed in core is significant in the context of the second hypothesis. They report fractures formed subparallel to S_{H} in Early Carboniferous limestones at greater than about 3000 m depth below the surface in North Dakota. Although they think that fluid pressures were normal-hydrostatic immediately after fracturing they do not rule out the possibility that fluid pressures could have been abnormally high during fracturing. Thus, jointing may occur simultaneously in both a thin near-surface channel and at greater depth. 'Dry' unloading joints form close to the Earth's surface where lateral relief is possible, and 'wet' tectonic joints, partly driven by high fluid pressures related to tectonic compaction, form at greater depths. Because shallow-formed joints may be exposed by modest denudation not long after their formation, stress directions inferred from them will generally be similar to those of the present day. More deeply formed, hydraulically driven, tectonic joints (possibly accompanied by parallel veins) will be exposed only after substantial exhumation.

The uniformity of strike of Ebro basin neotectonic joints throughout several plateaux and mesas bounded by slopes of different attitude indicates that local topography is not an important control. Discriminating between the hypotheses of control by lithology or altitude is not easy in the Ebro basin because the rocks are flat-lying and chalky limestones crop out only beneath the upper slopes of mesas. A regional survey of relative neotectonic joint development in relation to sedimentary facies and altitude in the basin is required.

6. Conclusions

1. Exposed neotectonic joints can be used to track approximately the horizontal axes of the contemporary stress field. The joints strike either parallel to the direction of the greatest horizontal stress or they symmetrically enclose a small angle about it. In both S. England/N. France and the Ebro basin (N. Spain), contemporary stress axes were successfully predicted from neotectonic joints before geophysical data became available.

2. The strikes of exposed neotectonic joint sets are generally uniform throughout areas exceeding 10000 km^2, reflecting the uniformity of orientation of the stress field to which they are related.

3. Abnormally high fluid pressures are not necessary to propagate neotectonic joints at very shallow (less than 500 m) crustal levels, but unloading consequent on denudation, and lateral relief resulting from uplift are, however, necessary prerequisites.

4. The majority of exposed neotectonic joints are extension fractures but a small proportion are hybrid-shear fractures. Differential stresses during neotectonic jointing are small and S_H is horizontal.

5. Neotectonic joints in the Ebro basin are best developed in a thin-near surface channel containing chalky limestones. The preferential development of joints in this channel is related to either the presence of weak but brittle limestones or uplift having raised the rocks to a height where lateral elongation is possible.

A. G. Becher, J. P. Turner and L. Arlegui assisted in the field and in many other ways. B. J. McConnell carried out the point load tests, and P. C. England made valuable suggestions for improving the paper. The Royal Society, the University of Bristol, King Abdul Aziz City for Science and Technology, Shell International Petroleum Company and the NERC funded fieldwork.

References

Bevan, T. G. & Hancock, P. L. 1986 A Late Cenozoic regional mesofracture system in southern England and northern France. *J. geol. Soc. Lond.* **143**, 355–362.

Brereton, N. R. & Evans, C. J. 1987 Rock stress orientations in the United Kingdom from borehole breakouts. *Rep. reg. Geophys. Res. Group, Br. geol. Surv.* no. R6 87/14, pp. 1–36.

Bock, E. & Franklin, J. A. 1972 The point load strength test. *Int. J. Rock Mech. Min. Sci.* **9**, 669–697.

Engelder, T. 1982 Is there a genetic relationship between selected regional joints and contemporary stress within the lithosphere of North America? *Tectonics* **1**, 161–177.

Engelder, T. 1985 Loading paths to joint propagation during a tectonic cycle: an example from the Appalachian Plateau, U.S.A. *J. struct. Geol.* **7**, 459–476.

Garcia Prieto, F. J. & Simón-Gómez, J. L. 1986 El campo de fallas miocenas de la Bardena Negra (provs. de Navarra y Zaragoza). *Bol. geol. Min. España* **97**, 693–703.

Gonzalez de Vallejo, L. I., Serrano, A. A., Capote, R. & De Vicente, G. 1988 The state of stress in Spain and its assessment by empirical methods. In *Rock mechanics and power plants* (ed. M. Romana), pp. 165–172. Rotterdam: Balkema.

Hancock, P. L. 1985 Brittle microtectonics: principles and practice. *J. struct. Geol.* **7**, 437–457.

Hancock, P. L. 1986 Joint spectra. In *Geology in the real world. The Kingsley Dunham volume* (ed. I. Nichol & R. W. Nesbitt), pp. 155–164. London: Institution of Mining and Metallurgy.

Hancock, P. L. 1987 Neotectonic fractures formed during extension at shallow crustal depths. *Mem. geol. Soc. China (Taiwan)* **9**, 201–226.

Hancock, P. L. & Engelder, T. 1989 Neotectonic joints. *Bull. geol. Soc. Am.* **101**, 1197–1208.

Hancock, P. L., Al-Kadhi, A., Barka, A. A. & Bevan, T. G. 1987 Aspects of analysing brittle structures. *Anns Tectonicae* **1**, 5–19.

Hodgson, R. A. 1961 Regional study of jointing in Comb Ridge–Navajo Mountain area, Arizona and Utah. *Bull. geol. Soc. Am.* **45**, 1–38.

Holst, T. B. & Foote, G. R. 1981 Joint orientation in Devonian rocks in the northern portion of the lower peninsula of Michigan. *Bull. geol. Soc. Am.* **92**, 85–93.

Julivert, M. 1978 The areas of Alpine folded cover in the Iberian Meseta (Iberian Chain, Catalonides, etc.). In *Geological atlas of Alpine Europe and adjoining areas* (ed. M. Lemoine), pp. 93–112. Amsterdam: Elsevier.

Kulander, B. R., Barton, C. C. & Dean, S. L. 1979 The application of fractography to core and outcrop fracture investigations. *Tech. Rep. U.S. Dept. Energy.* Morgantown Energy Technology Center, **METC/SP-79/3**, 1–174.

Narr, W. & Burress, R. C. 1984 Origin of reservoir fractures in Little Knife Field, North Dakota. *Bull. Am. Ass. Petrol. Geol.* **68**, 1087–1100.

Pollard, D. D. & Aydin, A. 1988 Progress in understanding jointing over the past century. *Bull. geol. Soc. Am.* **100**, 1181–1204.

Price, N. J. & Cosgrove, J. W. 1990 *Analysis of geological structures.* (502 pages.) Cambridge University Press.

Simón-Gómez, J. L. 1989 Late Cenozoic stress field and fracturing in the Iberian Chain and Ebro Basin (Spain). *J. struct. Geol.* **11**, 285–294.

Stewart, I. S. & Hancock, P. L. 1990 Brecciation and fracturing within neotectonic normal fault zones in the Aegean region. In *Deformation mechanisms, rheology and tectonics* (ed. R. J. Knipe & E. H. Rutter) *Spec. Publ. geol. Soc. Lond.* **54**, 105–112.

Turner, J. P. & Hancock, P. L. 1990 Relationships between thrusting and joint systems in the Jaca thrust-top basin, Spanish Pyrenees. *J. struct. Geol.* **12**, 217–226.

Discussion

R. Maddock (*Geoscience Ltd, U.K.*). In the Ebro Basin example, how is extensional strain partitioned between the joints and the normal faults in percentage terms?

P. L. Hancock. Because the NNW-striking neotectonic joints in the Ebro basin can be shown, on the basis of field evidence, to be younger than the NNE-striking normal faults strain partitioning was not involved. Formation of the joints resulted in no more than 1% elongation. Although this strain is trivial the uniformity of the inferred ENE–WSW extension direction throughout an area of least 10000 km^2 permits confidence in the idea that the joints are reflections of a strain of regional significance.

M. L. Zoback (*US Geological Survey, Menlo Park, U.S.A.*). I never thought I would stand up in defence of joints, but I think Dr Hancock has made some really important observations regarding neotectonic joints and their relationship to the contemporary stress field. One thing that has always bothered me in the Appalachian Plateau is the question of the large stratigraphic window. If we look at Devonian rocks, why should they be recording the modern stress field when they have been sitting there so long?

P. L. Hancock. Dr Zoback's remarks about the potential of neotectonic joints for tracking contemporary stress trajectories are much appreciated. A simple answer to her question is that old rocks are just as capable of experiencing contemporary stresses as young rocks. Whether old rocks respond by developing new structures, or by reactivating pre-existing ones, depends on the orientations and characters of the old structures, and the orientations and magnitudes of the contemporary stresses. The influence of residual stresses that have been locked-in since an ancient deformation episode generally seems to be slight.

B. Skipp (*Soil Mechanics Associates, U.K.*). Has Dr Hancock made a close examination of the Crag in East Anglia? Can he comment upon reports there of large concentrations of joints which perhaps are more fault-related and furthermore has he examined flat lying caliche type deposits which I would have thought should also show something of the contemporary stress field?

P. L. Hancock. I have not seen the structures referred to and that were described by Balson & Humphreys (1986) as cutting Plio-Pleistocene Crags in East Anglia. However, judging from the photographs in their paper, the neotectonic joints cutting the Cretaceous chalks and Palaeogene sands and clays in S. England/N. France are unlike the fissures they describe. The neotectonic joints are closed, rather than open, fractures and, in Suffolk, they terminate upwards in London Clay at the unconformity beneath the Red Crag (Bevan & Hancock 1986).

I agree that some Quaternary caliches might contain systematic joints that record the influence of the contemporary stress field. Caliches dated as latest Pliocene–Pleistocene in eastern Arabia are cut by systematic joints of neotectonic trend but only within their basal parts. Furthermore, these joints are direct continuations of joints cutting underlying Miocene–early Pliocene bedrocks. Thus they might be reflection cracks rather than tectonic structures. Non-systematic joints are abundant in the east Arabian caliches but there is little correlation between their average strikes and those of systematic neotectonic joints cutting the underlying bedrocks.

C. Vita-Finzi (*University College, London, U.K.*). Could the window be narrowed even further by making a study of joint systems in Quaternary or even in Holocene rocks? You would then benefit from the tighter correlation with the contemporary stress field and from the knowledge that the structures formed at the surface and have not undergone reactivation.

P. L. Hancock. A thorough search for systematic joints in Quaternary, and in particular in Holocene, sedimentary rocks would be most worthwhile. Such rocks are commonly cut by faults and fissures in regions that are actively extending. Paradoxically, regionally extensive and regularly orientated joint sets are generally lacking in such young rocks. However, non-systematic, randomly organised joints commonly cut late Pleistocene and Holocene rocks. For example, cemented beach rock in the Corinth area, which is cut by normal faults, contains only non-systematic joints. Several colleagues have reported that they have seen systematic joints in a variety of mainly older Quaternary materials but detailed descriptions of them are few, a notable exception being Caputo's (1990) account of joints in the Thessaly area of Greece. One reason why systematic joints in late Quaternary rocks are relatively rare might be that they have not experienced a stress history that involves both loading during burial and unloading during denudation.

Additional references

Balson, P. S. & Humphreys, B. 1986 The nature and origin of fissures in the East Anglian Coralline and Red Crags. *J. Quat. Sci.* **1**, 13–19.

Caputo, R. 1990 Geological and structural study of the recent and active brittle deformation of the Neogene–Quaternary basins of Thessaly (central Greece). *Scientific Annals of the Geological Department of the Aristotle University of Thessaloniki* **12**, 1–252.

Palaeostress determinations from fault kinematics: application to the neotectonics of the Himalayas-Tibet and the Central Andes

By J. L. Mercier, E. Carey-Gailhardis and M. Sébrier

CNRS Laboratoire de Géophysique et Géodynamique Interne, Université Paris-Sud, Bât. 509, 91405 Orsay, France

This paper concerns the determination of the stress deviator from a set of striated faults or of focal mechanisms of earthquakes. Southern Tibet and the Cuzco region of the High Andes illustrate the applied methodology. Then, the spatial and temporal changes in the tectonic régimes of Tibet and Central Andes are briefly examined. The Quaternary–Present-day stress patterns in these two high plateaus favour models explaining the spatial changes in the state of stress at a large wavelength as a result of a compensated elevated topography. Major temporal changes in the tectonic régimes are attributed tentatively to uplift for the Tibetan plateau and, for the already elevated Andean Plateau, to changes in the magnitude of the boundary forces due to subduction processes.

1. Introduction

Knowledge of the spatial and temporal changes in the continental stress fields is of great importance in understanding the sources of what is called the 'tectonic stresses' (Zoback *et al.* 1989). Geological data contribute to the definition of the Recent and Present-day stress fields but in particular they permit one to demonstrate temporal changes in the state of stress over geological periods. Indeed, structural analyses concern deformations and therefore, to have access to stresses, there must exist a simple linear relation between the strain tensor and the stress tensor. Therefore, these structural analyses essentially concern small brittle deformations and assume the following hypotheses: (1) the analysed body of rock is physically homogeneous and isotropic and, if pre-fractured, is also mechanically isotropic, i.e. the orientation of the fracture planes is random; (2) displacements on the fault planes are small with respect to the fault lengths and there is no ductile deformation of the material, and thus no rotation of the fracture planes. In general, rocks are neither homogeneous nor isotropic but at the scale of bodies of rock, some tens of metres to some kilometres long, these assumptions are often valid and coherent results, obtained from numerous sites in a given region, often give a good approximation of the regional state of stress.

Habitually, fault orientations have been compared with the orientations of shear planes formed in a homogeneous and isotropic medium and have been related to the principal stress directions acting in this medium (Anderson 1951). But, because of the frequent occurrence of inherited fractures, the practical use of this method is very

Phil. Trans. R. Soc. Lond. A (1991) **337**, 41–52

Printed in Great Britain

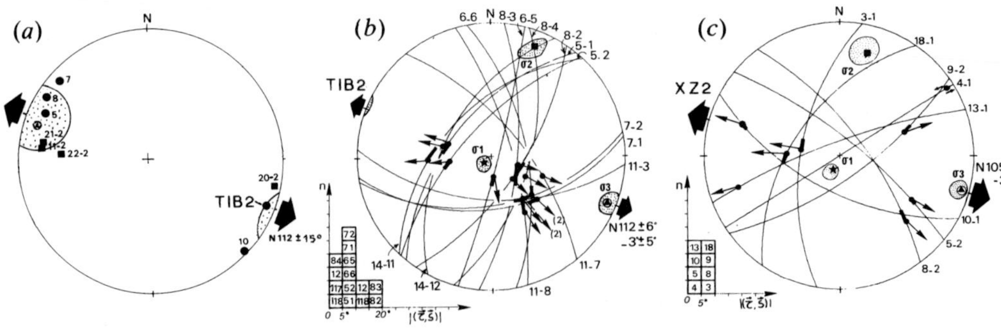

Figure 1. Quaternary and Present-day tensional (σ_3) directions in Tibet (figure 2) deduced from (*a*) recent minor normal faults. It shows Bingham statistics of the σ_3 axes obtained at different sites (Mercier *et al.* 1987), (*b*) major normal faults bordering the Yangbajin graben (TIB2), (*c*) focal mechanisms of teleseismic earthquakes (Carey-Gailhardis & Mercier 1987). Wulff stereonet, lower hemisphere, histograms give the deviations between the observed ($\boldsymbol{S}$) and predicted ($\boldsymbol{\tau}$) striations.

limited. Wallace (1951) and Bott (1959) have demonstrated that, if a pre-existing fault is reactivated, the orientation of the free slip on the fault plane depends on four parameters defining the principal stress reference and the stress ratio, $R = (\sigma_z - \sigma_x)/(\sigma_y - \sigma_x)$.

2. Palaeostress determination from fault kinematics in a fractured body of rock

(*a*) *Kinematics of faults measured in the field*

Using the mechanical approach of Wallace (1951) and Bott (1959), Carey & Brunier (1974) and Carey (1976) have proposed a quantitative computer-aided method to interpret the kinematics of a population of faults in a highly fractured body of rock. Their mechanical model assumes that the material is homogeneous and isotropic and that the deformations are small (see above) and, moreover, that (1) a tectonic event is characterized by a single homogeneous stress tensor, (2) the slip (responsible for the striation) occurs on each fault plane in the direction and sense of the resolved shear stress acting on the fault and (3) the slips are independent of each other. Then, deformation results from relative displacements of rigid blocks in an assemblage of blocks.

This model implies that the angle between the predicted $\boldsymbol{\tau}_i$ and the observed $\boldsymbol{S}_i$ slip-vectors is zero. The $\boldsymbol{\tau}_i$ components are a function of four parameters: the three Euler angles, which give the azimuthal directions of the orthogonal principal stress axes with respect to a geographical reference, and the relative ratio R of the principal deviatoric stress differences such that

$$R = (\sigma'_2 - \sigma'_1)/(\sigma'_3 - \sigma'_1)$$

(see demonstration in Carey (1976), and Carey-Gailhardis & Mercier (1987)). Numerous algorithms are now available to solve this problem. More sophisticated calculations are used to separate data corresponding to several states of stress. The reliability of the solutions obtained may be tested by calculation either of the deviations ($\boldsymbol{\tau}_i$, $\boldsymbol{S}_i$) on each fault plane or of the uncertainties in the principal stress directions (figure 1) and the R values.

(b) *Kinematics of faults deduced from focal mechanisms of teleseismic earthquakes*

The above numerical methods may also be applied to a population of focal mechanisms of teleseismic earthquakes, which have occurred in the same area, assuming a mean state of stress in the source region. This requires selection of the preferred seismic fault plane from each couple of nodal planes. This selection is made possible due to the fact that only one of the two slip vectors is the seismic slip in agreement with the principal stress axes. For this slip vector the R ratio, computed from Bott's (1959) formula, is such that $0 < R < 1$. Moreover, if one of the two nodal planes satisfies this condition, the other does not, except if the two nodal planes intersect each other along a principal stress axis (see Carey-Gailhardis & Mercier 1987). Generally, a population of focal mechanisms of major seismic events leads to a good evaluation of the regional state of stress in agreement with the state of stress deduced from Recent and active faulting (compare figure 1*a*, *b* and *c*).

(c) *Kinematics of faults deduced from focal mechanisms of aftershock sequences or of microseismic activity*

Focal mechanisms of aftershock sequences often show complex fault kinematics which result from heterogeneous deformation. Numerical analyses of such focal-mechanism data from regions submitted to an extensional tectonic régime (Mercier & Carey-Gailhardis 1989) clearly show a main group whose normal fault motions result from a tensional mean state of stress (TM) in good agreement with the regional state of stress deduced from recent Quaternary and active tectonics. The remaining complex fault motions are explained by compressional (TC) and/or tensional (TE) deviators. An important observation is that, within the uncertainties, these (TM), (TC) and (TE) deviators are coaxial. We suggest that in a highly fractured body of rock submitted to a state of stress, slip on many 'well-orientated' minor faults occurs so that motions of blocks statistically produced a deformation of the assemblage of blocks corresponding to the mean state of stress (TM). However, incompatibilities of block motions exist which induce internal elastic stresses in the blocks. These must be relaxed by slips on the block boundaries or by formation of new fractures. As these block displacements are constrained by the neighbouring blocks, it seems reasonable that these fault motions must be statistically compatible with the deformation of the whole assemblage and, thus, must be modelled statistically by deviators (TE, TC) having the same principal stress directions as those of the mean deviator (TM) but different values. Thus, provided some supplementary information about the active tectonic régime is known from other data, the (TM) deviator may be easily recognized. The (TE) and (TC) deviators model local slips and cannot be considered as other mean states of stress.

Detailed references concerning the numerical methods of stress determination from fault kinematics and earthquake focal mechanisms may be found in Zoback (1991), Zoback *et al.* (1989), Carey-Gailhardis & Mercier (1987) and Mercier & Carey-Gailhardis (1990).

(d) *Heterogeneous and homogeneous deformation of assemblages of rock blocks*

Thus data from the focal mechanisms of small earthquakes show that, in general, deformation of an assemblage of small blocks must be heterogeneous and that, in such a medium, motions of minor faults cannot be modelled by a single homogeneous stress deviator. However, it is a common observation that a population of striated

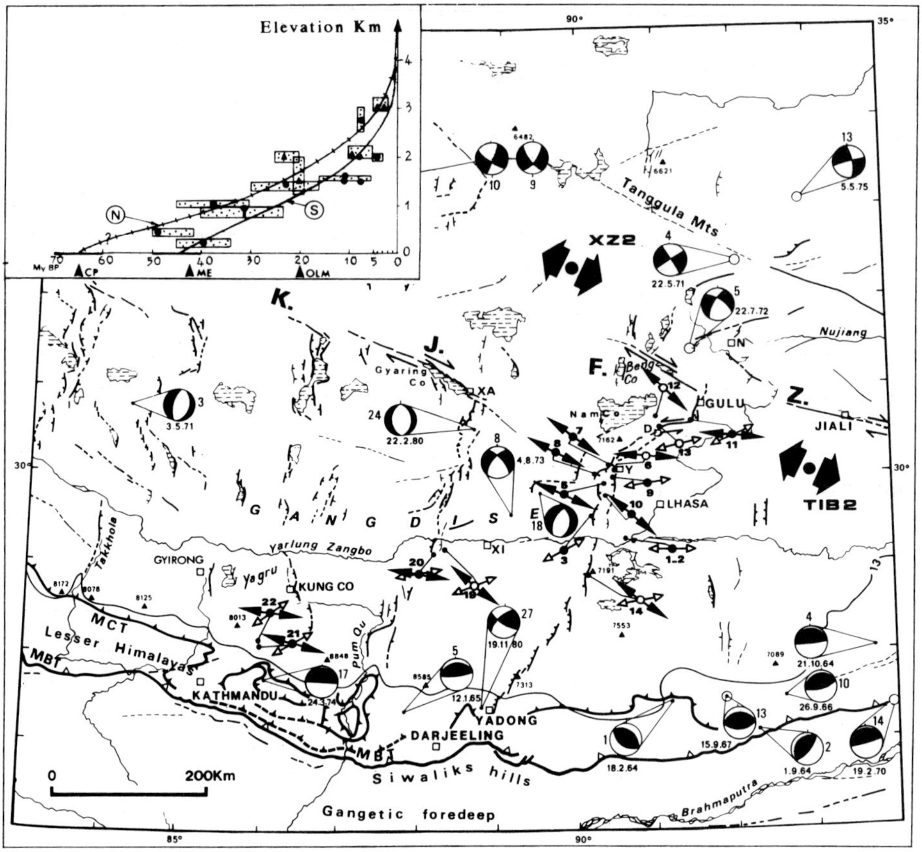

Figure 2. Seismotectonic map of Tibet (Mercier *et al.* 1987) showing the tensional directions deduced from analysis of recent minor faults (thin black arrows, figure 1*a*), major faults (large black arrows, TIB2, figure 1*b*) and focal mechanisms of teleseismic earthquakes (large black arrows, XZ2, figure 1*c*). Quaternary grabens are closely hachured. The inset (top left) schematically shows the Tibetan uplift inferred from variations of the vegetation in the Northern (N) and Southern (S) provinces.

faults, resulting from a single tectonic event, leads to the computation of a single stress deviator and this may appear surprising. It is noteworthy that numerous striations measured on faults in the field are mineral fibres and stylolitic striations which form by creep. In such a case, a pressure-dissolution process is active and annihilates the incompatibilities of block motions which should have existed as a result of stick–slip displacements.

3. Changes in the state of stress in Tibet during the Cenozoic and uplift of the high plateau

(*a*) *Analysis of fault kinematics in the field*

In Tibet geological mapping, satellite images and field observations (see references in Tapponnier *et al.* 1981 and Mercier *et al.* 1987) show that compressional structures striking E–W are cross-cut by N–S striking rift systems. Thus change in the tectonic régime is easily demonstrated. In the central part of the Indus–Zangbo suture, reverse faults and backthrusts affect Oligo-Miocene (*ca.* 30–15 Ma) conglomerates. Their kinematics are in agreement with a N–S to NNE–SSW direction of shortening (Mercier *et al.* 1987). The rift systems have been active probably since the Pliocene (*ca.* 5 Ma). Motions of the recent minor normal faults, some tens to several hundreds metres long, are in agreement with tensional σ_3 axes (figure 2) whose mean direction trends $112° \pm 15°$ (figure 1*a*). The major faults, several kilometres long, which border the Yangbajin graben (figure 2) also show kinematics which agree with a σ_3 axis trending $112° \pm 6°$ (figure 1b). Finally, inversion of the focal mechanisms of teleseismic earthquakes located in a large area (figure 1*c*) shows that these are in agreement with a tensional σ_3 axis trending $105° \pm 7°$.

(*b*) *The spatial change in the present-day state of stress of the Himalayas-Tibet*

Thus the Tibetan plateau (mean altitude *ca.* 5000 m) is subjected to a tensional (σ_3) axis trending $112° \pm 6°$. On the other hand the Himalayan lowlands (altitude 300–1350 m above sea level) are subjected to a NNE–SSW trending compression (σ_1) as demonstrated by superficial folds and thrusts affecting Plio-Pleistocene deposits and by focal mechanisms of earthquakes (figure 2). Therefore, the maximum horizontal stress ($\sigma_{\mathrm{H,\,max}}$) trajectory trends NNE–SSW, $022° \pm 6°$ in the high plateau, i..e roughly parallel to the India–Eurasia convergence in that central part of the India–Asia collision zone; $\sigma_{\mathrm{H,\,max}}$ is σ_1 in the lowlands and σ_2 in the highlands. This spatial change in the state of stress favours models that explain the occurrence of normal faulting in Tibet as a consequence of its high elevation (Tapponnier & Molnar 1976; England & McKenzie 1982). These model emphasize the importance of topographic features of large extent which are compensated at depth by mass deficiencies due to a thick crustal root and a light asthenospheric mantle below it which perturb the vertical lithospheric stress σ_{zz} (see Fleitout & Froidevaux 1982). In the Himalayan lowlands, tectonics are compressional, thus $\sigma_{\mathrm{H,\,max}}$ is σ_1 and σ_{zz} is σ_3. In regions of high elevation σ_{zz} increases with the topography. In the Tibetan plateau σ_{zz} exceeds $\sigma_{\mathrm{H,\,max}}$, thus σ_{zz} is σ_1, $\sigma_{\mathrm{H,\,max}}$ becomes σ_2 and $\sigma_{\mathrm{H,\,min}}$ trending WNW–ESE becomes σ_3 thereby allowing extension to occur in this direction, i.e. orthogonal to the compressional direction acting in the adjacent lowlands (figure 2).

(*c*) *Temporal changes in the Cenozoic state of stress and uplift of Tibet*

Timing of the Tibetan uplift is supported by geological data provided essentially by palynological assemblages and coal beds (figure 2). These floral assemblages depend on the elevation above sea level but during the Pliocene–Quaternary overall climatic cooling has increased the effect of uplift on the floral environment. In spite of these uncertainties, it clearly appears that compressional structures of Upper Cretaceous–Palaeocene to Oligo-Miocene age formed in Tibet when its elevation was below an altitude of 2000 m. On the other hand, extension started during the Pliocene when the Tibetan plateau reached an elevation probably exceeding 3000 m. This observed temporal change in the state of stress gives additional support to the model that extension in the Tibetan plateau is a consequence of its high topography.

5. Changes in the state of stress in an already elevated plateau, the Andean plateau as an example

In the High Andes, the successive tectonic régimes during the Pliocene–Pleistocene are deduced from a detailed study of the successive fault kinematics. The Cuzco fault

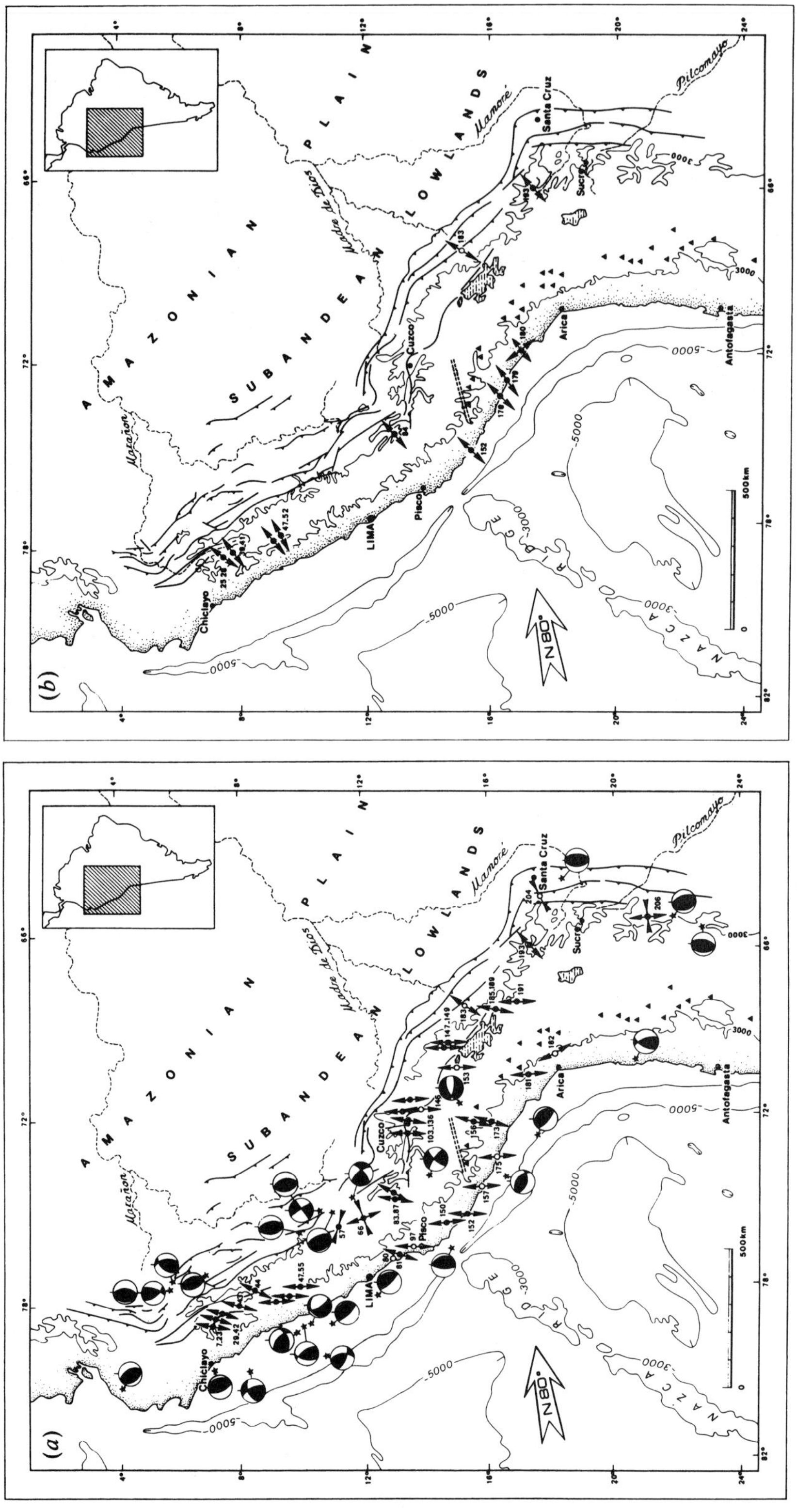

Figure 3. Principal stress directions deduced from analysis of fault kinematics in Peru and Bolivia. (*a*) Pleistocene and active faults and focal mechanisms of earthquakes, (*b*) NE–SW trending tensional σ_3 axes of latest Miocene–Pliocene age. Site numbers are those of Mercier *et al.*'s (1991) paper.

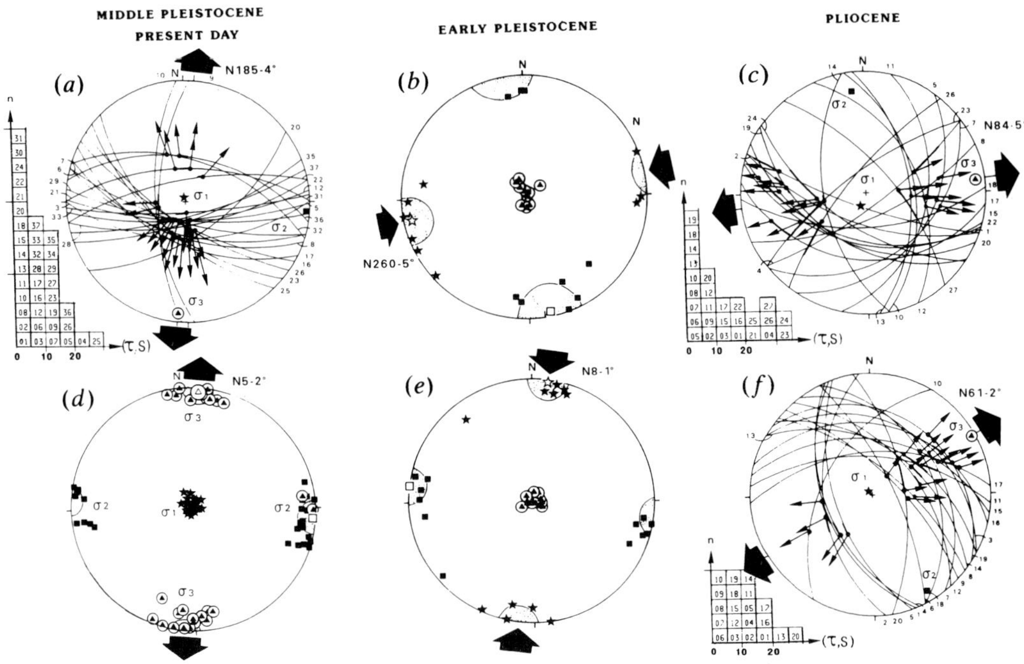

Figure 4. Fault slip-vector data and principal stress directions of the Cuzco region of (*a*), (*d*) Mid-Pleistocene to Present day, (*b*), (*e*) Early Pleistocene, (*c*), (*f*) and Pliocene age. Same symbols as on figure 1.

system (sites 103–109, figure 3*a*) in Peru is chosen as an example to illustrate the methodology used (see data in Sébrier *et al.* 1985, 1988; Mercier *et al.* 1991).

(*a*) *Analysis of fault kinematics in the field*

Minor faults have been analysed in the vicinity of the major fault system and in the adjacent grabens. Normal faults affecting Early Pleistocene (*ca.* 2–1 Ma) and Holocene (< 10 ka) deposits show kinematics which are in agreement with N–S trending tensional (σ_3) axes whose mean direction trends $005° \pm 6°$ (figure 4*d*). Compressional deformations also affect these Early Pleistocene deposits but they predate the deposition of terraces of Mid-Pleistocene age (*ca.* 0.4 Ma). They are synsedimentary and postsedimentary folds and reverse faults which are often offset by the above normal faults; normal striations superimposed upon reverse striations also are often observed on reverse fault planes. The kinematics of both synsedimentary and postsedimentary reverse faults are in agreement with either E–W or N–S compressional (σ_1) axes whose mean directions trend $260° \pm 10°$ and $008° \pm 7°$, respectively (figure 4*b*, *e*). An older normal faulting is coeval with sedimentation of lacustine and volcanic deposits. It postdates folding of probably latest Miocene age (*ca.* 7 Ma) and predates the above Early Pleistocene compression. This suggests that this older extension is of Pliocene age (*ca.* 5–2 Ma) which is clearly demonstrated in other regions of Peru. Kinematics of these synsedimentary faults are in agreement with tensional axes trending 060–090° (figure 4*f*).

The major fault planes affecting recent deposits, and observed in trenches dug across the seismic or Holocene fault 'scarplets', show a single direction of motion which is in agreement with a 185° trending σ_3 axis (figure 4*a*). However, the major

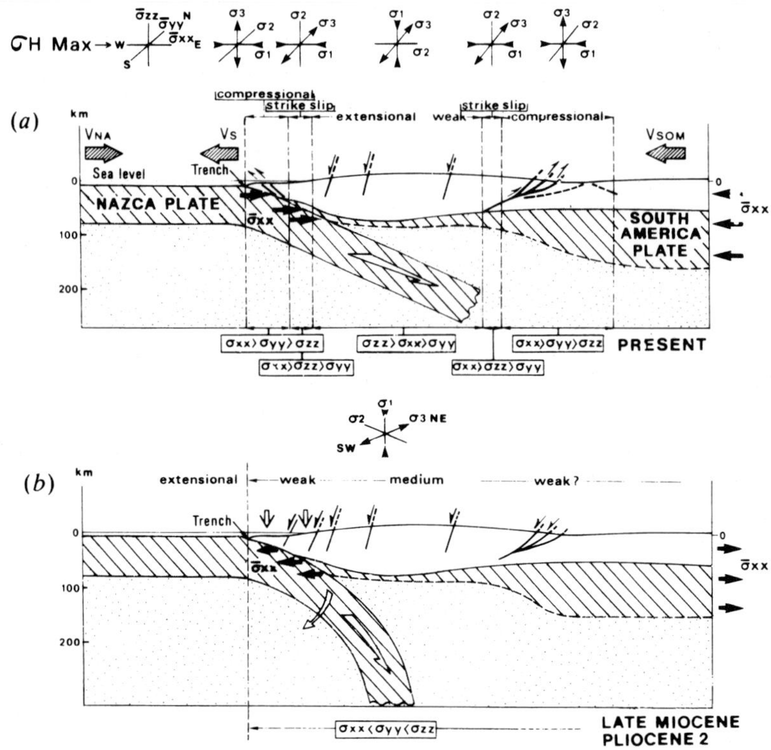

Figure 5. Suggested qualitative scenarios to interpret the change in the Andean state of stress (Mercier *et al.* 1991). (*a*) Present-day situation, σ_{xx} parallel to the convergence is $\sigma_{H,\max}$, N–S trending extension in the High Andes is an effect of compensated high topography (figure 3*a*); (*b*) Late Miocene–Pliocene situation, σ_{xx} is $\sigma_{H,\min}$ possibly as a result of slab retreat; a WNW–ESE extension is active, possibly throughout the whole Andes.

fault planes affecting the bed-rock show several superimposed families of striations. The younger striations are normal and agree with the above 185° trending tension. They postdate reverse and reverse strike–slip striations which belong to two different groups; one is in agreement with an E–W and the other with a N–S compression. The oldest striations are normal and are in agreement with a 085° trending σ_3 axis (figure 4*c*).

In conclusion, the successive kinematics of the major faults, when compared with the successive dated kinematics of the minor faults, permit one to establish a sequence of regional states of stress. In the High Andes this is as follows: (1) a Pliocene extensional tectonic régime having an NE–SW to E–W σ_3 axis, (2) an Early Pleistocene compressional tectonic régime showing two families of fault kinematics resulting from either an E–W or a N–S compression (σ_1) and (3) a Mid-Pleistocene–Present-day extensional régime having a roughly N–S σ_3 axis.

(*b*) *The spatial changes in the Present-day state of stress in the Central Andes*

A model of compensated elevated topography (figure 5*a*) is also convenient to explain the spatial change in the stress field in the Central Andes. A roughly E–W compression is active in the sub-Andean lowlands, along the Amazonian plain, and at the contact between the Nazca and South American plates and a roughly N–S extension is active in the High Andean plateau (figure 3*a*). Extensional tectonics occurs at a mean elevation of *ca.* 3000 m, lower than in Tibet. This difference reflects the magnitudes of the compressive forces related to the convergence which are lower in the Andean subduction than in the Himalayan collision. In Peru, the tensional direction trends more NNE than expected from the 085° direction of convergence between the two plates.

(*c*) *The temporal changes in the state of stress in the High Andes during the Pliocene–Pleistocene*

The detailed analysis of temporal changes in the state of stress in the High Andes during this period is beyond the scope of this paper. Two important points must be emphasized: (1) the elevation of the Andes above sea level has not significantly changed during the past 5 Ma; (2) palaeomagnetic data show that the rotation of material in the Central Andes has been of the order of *ca.* 15° since the latest Miocene and cannot explain the differences between the σ_3 direction of Quaternary (figure 3*a*) and of Pliocene (figure 3*b*) age. Therefore, if the rheology of the Andean lithosphere has not changed significantly during this period, variations in magnitude of the boundary forces (figure 5*a*, *b*) are necessary to explain the changes in the tectonic régimes (figure 3*a*, *b*) during this period (see Mercier *et al.* 1991). These variations should be related to the subduction processes along the Andean margin (see, among others, Bott *et al.* 1989).

6. Concluding remarks

1. Many algorithms are now available to compute stress tensors from populations of faults or from earthquake focal mechanisms. The quality of the solutions obtained depends on the isotropy of the fault directions and on the deviations ($\boldsymbol{\tau}$, $\boldsymbol{S}$) between the observed ($\boldsymbol{S}$) and predicted ($\boldsymbol{\tau}$) slip vectors. Therefore, fault data should be reported on stereonets. These data must be clear enough to eventually permit digitization of the data. The histogram of the deviations ($\boldsymbol{\tau}$, $\boldsymbol{S}$) or the uncertainties in the stress directions (figure 1*b*, *c*) must be given also. In the database, this may be indicated by the mean and standard values of the deviations ($\boldsymbol{\tau}$, $\boldsymbol{S}$).

2. Concerning the palaeostress determination, as several deviators computed from a population of striated faults do not necessarily mean several actual states of stress (see §2*c*) it is of primary importance to indicate the geological data which have permitted one to separate the different tectonic régimes. Superimposed striations should be indicated on the stereographic projection of the faults with their relative chronology when observed. In the database the relative chronology of different families of striations at a given site, which lead to the computation of several stress tensors, should be indicated. The age of the formations affected by a family of striations and the age of the overlying formation, which is not affected by this family, should also be included in the database. Finally, it is useful to give the statistical mean direction of stress axes (σ_1 or σ_3 according to compressional or extensional régimes) obtained from minor fault sites and, if possible, to give the corresponding axis obtained from major faults in the same area (figure 4*a*, *d*).

3. Palaeostress determinations demonstrate temporal changes in the tectonic régimes and in the stress orientations which clearly show that the observed Present-day state of stress may not be extrapolated in general over long geological periods when formulating geodynamic reconstructions. The palaeostress-field analysis should be associated with palaeomagnetic studies to eventually demonstrate rotations of the stress directions 'frozen' in the material. Palaeostress fields are good indicators of

changes in the geodynamic conditions (see, among others, Philip 1987; Le Pichon *et al.* 1988; Zoback *et al.* 1991; Mercier *et al.* 1991).

This work has been supported by the programme DBT 'Instabilités' (contribution CNAS-INSU-DBT no. 351), Institut National des Sciences de l'Univers, Paris.

References

* Detailed references concerning the numerical methods of stress determinations are found in these references.

Anderson, E. M. 1951 *The dynamics of faulting*. Edinburgh: Oliver & Boyd.

Bott, M. H. P. 1959 The mechanics of oblique slip faulting. *Geol. Mag.* **96**, 109–117.

Bott, M. H. P., Waghorn, G. D. & Whittaker, A. 1989 Plate boundary forces at subduction zones and trench-arc compression. *Tectonophys.* **170**, 1–15.

Carey, E. 1976 Analyse numérique d'un modèle mécanique élémentaire appliqué à l'étude d'une population de failles: calcul d'un tenseur moyen des contraintes à partir des stries de glissement. *Thèse Univ. Paris-Sud, Orsay* (167 pages).

Carey, E. & Brunier, B. 1974 Analyse théorique et numérique d'un modèle mécanique élémentaire appliqué à l'étude d'une population de failles. *C.r. hebd. Séanc. Acad. Sci., Paris* **279**, 891–894.

*Carey-Gailhardis, E. & Mercier, J. L. 1987 A numerical method for determining the state of stress using focal mechanisms of earthquake populations. *Earth planet. Sci. Lett.* **82**, 165–179.

England, P. C. & McKenzie, D. P. 1982 A thin viscous sheet model for continental deformation. *Geophys. Jl R. astr. Soc.* **70**, 295–321.

Fleitout, L. & Froidevaux, C. 1982 Tectonics and topography for a lithosphere containing densities heterogeneities. *Tectonics* **1**, 21–56.

Le Pichon, X., Bergerat, F. & Roulet, M. J. 1988 Plate tectonics and tectonics leading to the Alpine belt formation, a new analysis. *Geol. Soc. Am.* **218**, 111–131.

Mercier, J. L., Armijo, R., Tapponnier, P., Carey-Gailhardis, E. & Han, T. L. 1987 Change from Late Tertiary compression to Quaternary extension in southern Tibet during the India–Asia collision. *Tectonics* **6**, 275–304.

*Mercier, J. L. & Carey-Gailhardis, E. 1989 Regional state of stress and characteristic fault kinematics instabilities shown by aftershock sequences: the aftershock sequences of the 1978 Thessaloniki (Greece) and 1980 Campania-Lucania (Italia) earthquakes as examples. *Earth planet. Sci. Lett.* **92**, 247–264.

Mercier, J. L., Sébrier, M., Lavenu, A., Cabrera, J., Bellier, O., Dumont, J. F. & Machare, J. 1991 Changes in the tectonic regime above a subduction zone of Andean Type: the Andes of Peru and Bolivia during the Pliocene-Pleistocene. *J. geophys. Res.* (In the press.)

Philip, H. 1987 Plio-Quaternary evolution of the stress field in Mediterranean zones of subduction and collision. *Annls Geophysicae* **87.03B**, 301–309.

Tapponnier, P. *et al.* 1981 The Tibetan side of the India–Eurasia collision. *Nature, Lond.* **294**, 405–410.

Tapponnier, P. & Molnar, P. 1976 Slip-line field theory and large-scale continental tectonics. *Nature, Lond.* **264**, 319–324.

Sébrier, M., Mercier, J. L., Mégard, F., Laubasher, G. & Carey-Gailhardis 1985 Quaternary normal and reverse faulting and the state of stress in Central Andes of South Peru. *Tectonics* **4**, 739–780.

Sébrier, M., Mercier, J. L., Machare, J., Bonnot, D., Cabrera, J. & Blanc, J. L. 1988 The state of stress in an overriding plate situated above a flat slab: the Andes of Central Peru. *Tectonics* **7**, 895–928.

Wallace, R. E. 1951 Geometry of shearing stress and relation to faulting. *J. Geol.* **59**, 118–130.

Zoback, M. L. 1991 First and second order patterns of stress in the lithosphere, World stress Map Project. *J. geophys. Res.* (In the press.)

*Zoback, M. L. *et al.* (29 authors) 1989 Global patterns of tectonic stresses. *Nature, Lond.* **341**, 291–298.

Discussion

S. Stein (*Northwestern University, Evanston, U.S.A.*). When the motion across the Peru–Chile Trench is examined from space geodetic data, that's the motion between Easter Island and Arequipa, about 15 mm a^{-1} of the motion that one would expect between the Nazca Plate and the central interior of South America is missing. What is happening?

J. L. Mercier. Indeed the motion between Easter Island and Arequipa shown by geodetic data appears to be anomalous and we have no certain explanation to propose. However, taking into account the fact that the sub-Andean shortening is about ten times higher along the Bolivian orocline (14–19° S) than further north, i.e. along the Huancabamba traverse (5° S), one might expect a slower convergence rate in the area of the Arica elbow (18° S).

P. Hancock (*Bristol University, U.K.*). Fault striation analysis is a very powerful and elegant technique, but there is just one assumption that I have never understood and that is that the striation on all faults will be parallel to the resolved shear stress. This is a reasonable assumption, if we assume that all the faults are independent elliptical surfaces, but if they interact, as we know many real fault systems do, with perhaps two or three faults interacting, the sense of slip will be partly determined by the shape of the block and won't always be parallel to that resolved shear stress.

J. L. Mercier. It is a real fact that in a highly fractured body of rock fault motions are not independent. Focal mechanisms of aftershock sequences and microseismic activity show that deformation of an assemblage of small blocks is heterogeneous and cannot be modelled by a single homogeneous stress deviator. This results from incompatibilities of motions of neighbouring blocks. Yet it is noteworthy that numerous fault striations measured in the field are mineral fibres and stylolitic striations which form by creep. In these cases a pressure-dissolution process has annihilated the incompatibilities of fault motions. I had no time to explain this during my talk; it concerns §§2*c* and *d* of our paper.

P. England (*Oxford University, U.K.*). If both horizontal principal stresses are tensional, why are there strike–slip faults in the Himalayas?

J. L. Mercier. Supposing that $\sigma_{\mathrm{H,max}}$, due to the convergence, and σ_{zz} remain constant, the deviatoric σ_2 value in the high plateau depends on the E–W boundary forces. Because the southern part of the plateau, between the Himalayan syntaxes (see Mercier *et al.* 1987, fig. 12), does not have the possibility to 'flow' in the E–W direction, the deviatoric σ_2 value must be weakly compressional or extensional (stress ratio $R \geqslant 0.45$). Because the northern part of the plateau does have the possibility to extend in an E–W direction, the deviatoric σ_2 value must be compressional ($R < 0.5$). Numerical simulations show that if the σ_2 value is extensional, only nearly vertical faults (except those normal to σ_3) may have a strike–slip motion; other faults, whatever their strikes, have high normal components of motion (see the stereonet in figure 1*b* of recent faults in Southern Tibet; $R = 0.46 \pm 0.14$). If the σ_2 deviatoric value is compressional, faults having dips less than 70° may have a strike–slip motion, especially those striking 060–070° and 150–160°

(see stereonet in figure 1*c* of seismic faults from focal mechanisms of Northern Tibet earthquakes, $R = 0.17 \pm 0.07$). Differences in fault motions are seen in these two cases when one compares the motions of the faults labelled 11-7 (figure 1*b*) and 8-2 (figure 1*c*) which have roughly the same strike and dip (this problem has been discussed in Mercier *et al.* (1987), §4.3, fig. 12).

Seismological constraints on stress in the oceanic lithosphere

By Seth Stein and Aristeo Pelayo
Department of Geological Sciences, Northwestern University, Evanston, Illinois 60208, U.S.A.

Studies of earthquakes provide some of the basic constraints on the stress within the oceanic lithosphere. The focal mechanisms of earthquakes indicate the directions of the principal stress axes. These directions can be compared with the stresses predicted by plate driving-force models, given the caveat that they can be biased by the locations of pre-existing weak zones and local processes. One basic result, that lithosphere older than about 35 Ma is in deviatoric compression, can constrain the basal drag on plates and hence mantle viscosity. A second result, the absence of a broad zone of ridge-normal extension in younger lithosphere, suggests that the ridge axis is rheologically weak. The focal depths of earthquakes, which indicate the depth to which the lithosphere is strong enough to support seismicity, can be combined with thermal models and laboratory rock mechanics to constrain the rheology of the lithosphere, both in plate interiors and at plate boundaries. An additional data type, which could in principle yield inferences about the magnitudes of stress, consists of the seismic moments, magnitudes, and source time functions of the earthquakes. Some initial results suggest that transform fault earthquakes release seismic energy more slowly than ridge events, perhaps due to lower stress drops or source geometry.

1. Introduction

For tectonic purposes, ideally we would like to know the directions and magnitudes of the principal stress components as a function of position and depth in the lithosphere. Such data, however, can only be derived from overcoring measurements in mines or at the Earth's surface and hence are sparse. Hence much of the available information about stress in the lithosphere is derived from other techniques. Among these are observations of earthquakes, which are a manifestation of the stresses and thus a source of information about them. For example, earthquakes provide 50% of the data in a recent global compilation of stress orientations (Zoback *et al.* 1989). In oceanic lithosphere, where borehole data are sparse, earthquakes provide almost all of the available data.

Here, we review some inferences about stress in the oceanic lithosphere that can be derived from earthquake data. The oceanic lithosphere includes material within plate interiors, and that at the boundaries: the newly formed material at spreading centres, that being subducted, and that undergoing strike–slip motion at transforms. Earthquakes at the plate boundaries reflect the forces and stresses there which presumably have a large effect on the stresses, and hence earthquakes, within the plates. Here, we focus primarily on oceanic intraplate earthquakes. These earthquakes (figure 1), though few compared with plate-boundary earthquakes,

Phil. Trans. R. Soc. Lond. A (1991) **337**, 53–72
Printed in Great Britain

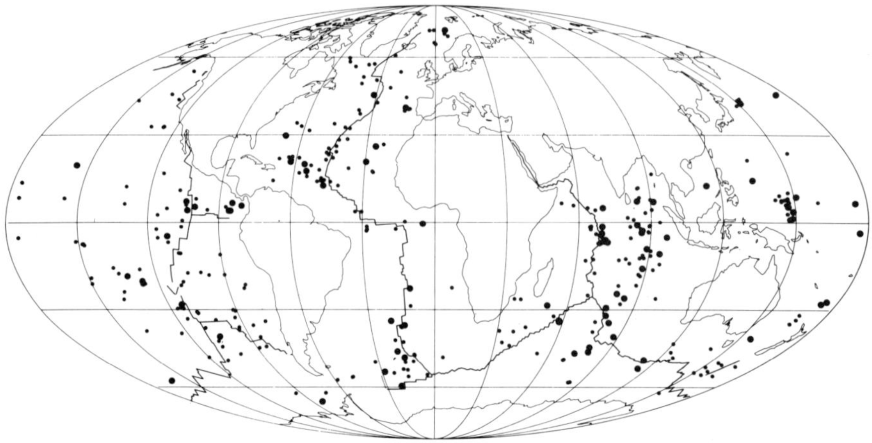

Figure 1. Locations of oceanic intraplate earthquakes between 1964 and 1983. Larger symbols show earthquakes with $m_b \geqslant 5.4$. Events of this size are generally large enough for mechanism and depth study (Bergman 1986).

Table 1. *Seismological constraints on lithospheric stress*

observable	constraints	caveats
focal mechanism	principal stress directions	can be biased by pre-existing faults
focal depth	strength against depth	interpretation requires thermal, rheological, and faulting models
source parameters (magnitudes, moment, time function)	stress drop	interpretation requires assumptions on fault geometry, rupture process, and relation to ambient stress

convey some interesting information about intraplate stresses and plate driving forces. The oceanic plates should be those most easily understood from a driving force standpoint, since they are presently the fastest moving plates and should be more uniform thermally and mechanically, and thus less affected by local stress perturbations from density inhomogeneities than the continents.

The inferences that can be drawn from oceanic intraplate earthquakes can be divided by the type of observations used. Table 1 shows the three types of observations discussed here, the constraints each provides, and some of the limitations of the resulting inferences. The locations of earthquakes and their focal mechanisms provide constraints to test and discriminate models of plate driving forces. In this application, the stresses inferred from the focal mechanisms are compared with those predicted by various models. The focal depths of earthquakes constrain the stresses by indicating depths at which the lithosphere supports stresses adequate for seismic failure. A third data type, which might yield inferences about the magnitudes of stress, consists of source parameters which describe the rupture process and may provide information on the stress drop during the earthquake.

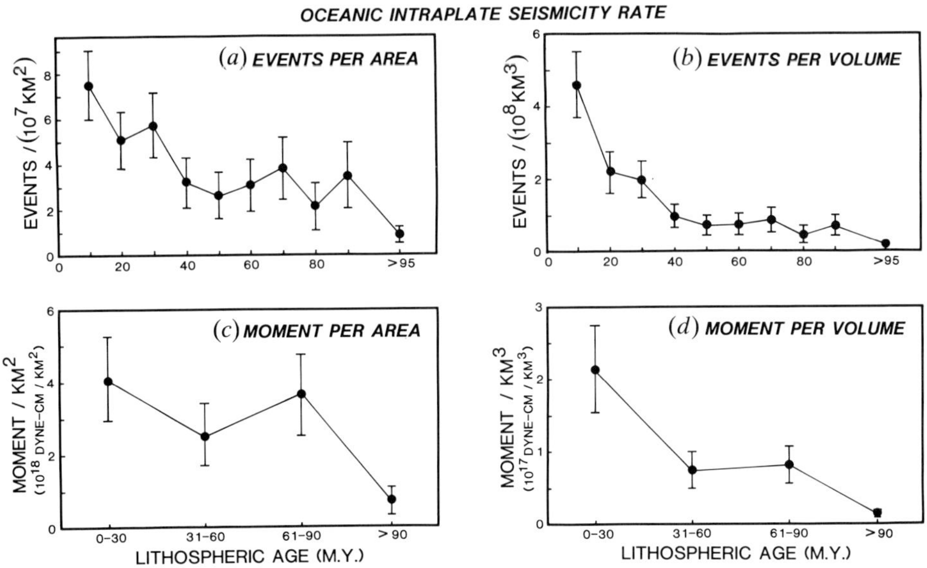

Figure 2. (*a*), (*b*). Number of oceanic intraplate earthquakes with $m_b \geqslant 5.0$ per surface area (*a*) and per volume of lithosphere above the 750 °C isotherm (*b*) as a function of lithospheric age. Both the number of earthquakes and the cumulative seismic moment (*c*, *d*) decrease with age. The area/age relations are from Parsons (1982), and the earthquake data are from 1964 to 1979 (Wiens & Stein 1983).

2. Focal mechanisms

An important application of earthquake seismology is to determine the geometry of the fault giving rise to the observed seismic radiation. For earthquakes of magnitude greater than about 5.5, it is generally possible to determine two orthogonal nodal planes, one of which is the fault plane (Aki & Richards 1980). The observed fault geometry is related to the principal stress axes via the simple Coulomb criterion (Jaeger & Cook 1976), which predicts fracture on a plane of maximum shear stress. These planes are halfway between the maximum and minimum principal stress directions, and include the intermediate principal stress direction. Since the shear stresses acting on the Earth's free surface are zero, near the surface one principal stress axis is vertical, and the other two are horizontal. Using this approach, the directions and relative magnitudes of the principal stresses are inferred from the fault geometry.

Implicit in this approach is the idea that the fault formed in response to the stress field, such that its geometry reflects the stress. Often, however, the earthquake occurred on a pre-existing weak zone or fault. In this case, inferring principal stress directions 45° from the fault is inappropriate, since a range of principal stress directions will give rise to slip on the pre-existing fault, rather than formation of a new fault (Handin 1969). Hence stress directions inferred from individual mechanisms may reflect the orientations of weak zones rather than regional tectonic stresses (McKenzie 1969*a*; Stein 1979). Ideally, the regional stress field would be estimated from a number of events in a given area showing a consistent inferred stress orientation. Unfortunately, intraplate earthquakes are rare enough that an adequate sample of earthquakes is generally not available for most regions, and inferences are drawn from single events.

The primary application of the stresses inferred from mechanisms is in the development and testing of models of the forces giving rise to regional stress fields which vary smoothly over large areas. Of special interest are the stresses expected to result from the forces driving-plate motions (Harper 1975; Forsyth & Uyeda 1975; Richter 1977; Richter & McKenzie 1978). Various studies (Solomon *et al.* 1975; Mendiguren & Richter 1978; Richardson *et al.* 1976, 1979; Richardson & Cox 1984; Wortel & Cloetingh 1985; Cloetingh & Wortel 1985, 1986) thus compare the inferred stresses to those predicted by models. A complication in this approach is the possibility that relatively localized stress sources can also give rise to seismicity, such as is observed in young oceanic lithosphere (discussed shortly) or on passive continental margins (Stein *et al.* 1989).

To develop some basic ideas, it is useful to consider first the variation of seismicity and mechanism types with lithospheric age. Comparison with age is a natural starting point, because the thermal evolution of the oceanic lithosphere is expected to give rise to the driving forces. This approach attempts to identify general patterns, while reducing the dependence on individual earthquakes, which pose the difficulties discussed earlier.

An interesting observation is that the level of seismicity, in terms of both the number of earthquakes and the seismic moment release, decreases dramatically with lithospheric age (figure 2). This effect is especially noticeable when corrected for the observation (discussed later) that the maximum depth of earthquakes appears to be controlled by temperature, so that the relevant variable is the volume of lithosphere of a given age above a limiting isotherm. This may provide a useful constraint if, as often implicitly assumed, the amount of seismicity is related approximately to the magnitude of stresses. The focal mechanisms of the earthquakes also vary with age (figure 3). Lithosphere older than about 35 Ma is in deviatoric compression, as shown by the thrust and strike–slip mechanisms. Younger lithosphere shows both extension and compression, with the former especially common in the Indian Ocean.

Models that seek to explain the seismological observations attempt to relate the stresses inferred from seismicity to those predicted by plate driving-force models. The models postulate a number of forces which contribute to the stress in any region. The stress at any location can be modelled by numerically finding the net effect of these forces for the entire plate.

Various authors have attempted to estimate the relative importance of different forces. For simplicity, we consider only three here. One, misleadingly termed 'ridge push', should arise from density gradients due to the cooling of the lithosphere. Within a region of the lithosphere, ridge push can be thought of as a body force, which is zero at the ridge and increases linearly with age, acting in the direction of the age gradient (Lister 1975; Parsons & Richter 1980). This force should give rise to deviatoric compression and hence thrust fault mechanisms. In young lithosphere the compression would be oriented approximately in the spreading direction, whereas for older lithosphere, the overall plate geometry can cause the compression direction to deviate from the local age gradient. A second major force, slab pull, should result from the density anomaly of the cold subducting lithospheric slabs (McKenzie 1969*b*). This force, which would depend on the thermal structure of the slab, varies with the age of the subducting lithosphere. A third force should arise from drag on the base of the plates, which resists their motion over the viscous mantle. At any point, this force should be proportional to, and directed opposite to, the absolute plate velocity (Solomon & Sleep 1974).

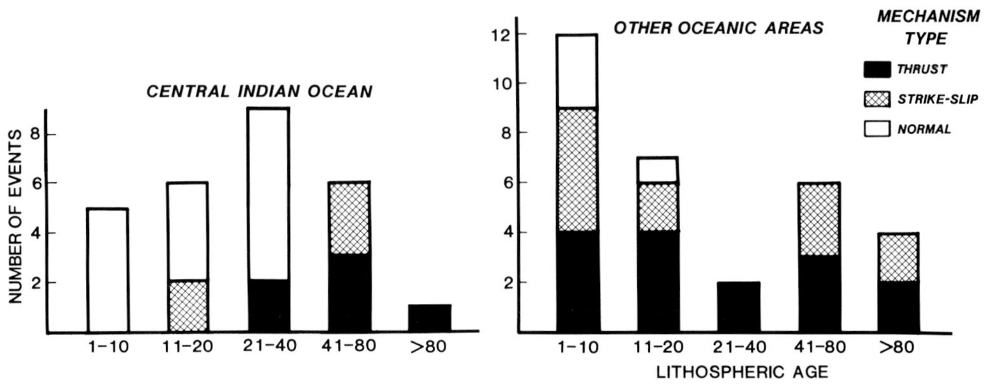

Figure 3. Mechanism type as a function of lithospheric age for oceanic intraplate earthquakes. Older oceanic lithosphere is in compression, whereas younger lithosphere has both extensional and compressional mechanisms. Extensional events are located primarily in the Central Indian Ocean (Wiens & Stein 1984).

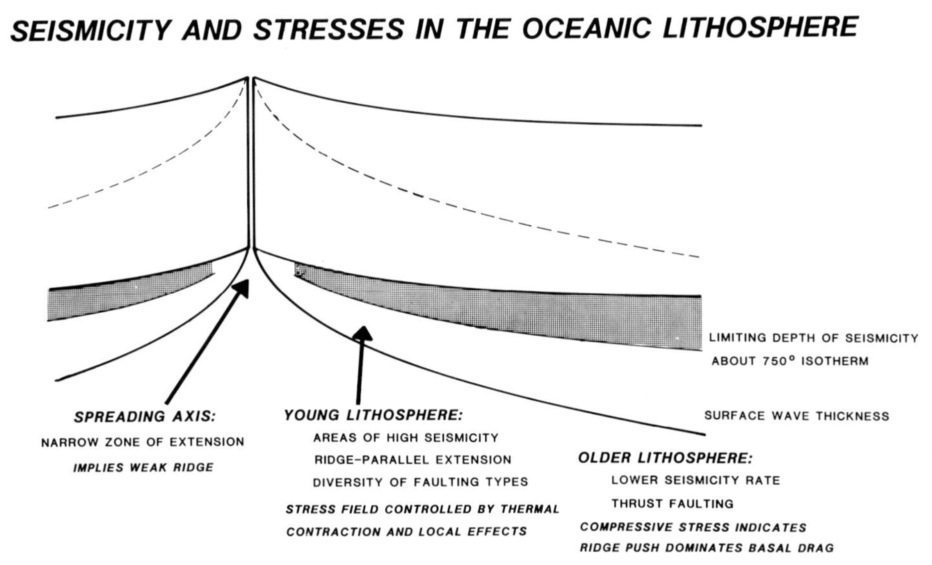

Figure 4. Schematic diagram relating seismicity observations to the tectonics of the oceanic lithosphere (Wiens & Stein 1985).

Following this approach, a number of studies have considered the relation of the focal mechanism data to the stresses predicted by different driving-force models. Since space precludes discussion of all of these, we focus on several key ideas, summarized in figure 4. Specifically, we summarize some ideas about the stresses firstly in old lithosphere, and then secondly in young lithosphere.

3. Focal mechanisms and stresses in old lithosphere

The observation of thrust faulting in older lithosphere has for some time been attributed to the compression expected from ridge push (Mendiguren 1971; Forsyth 1973; Sykes & Sbar 1974). The platewide models listed earlier generally predict this compression, although the direction at any point is not always that predicted. Given the difficulties with inferring stress directions from isolated mechanisms, a misfit

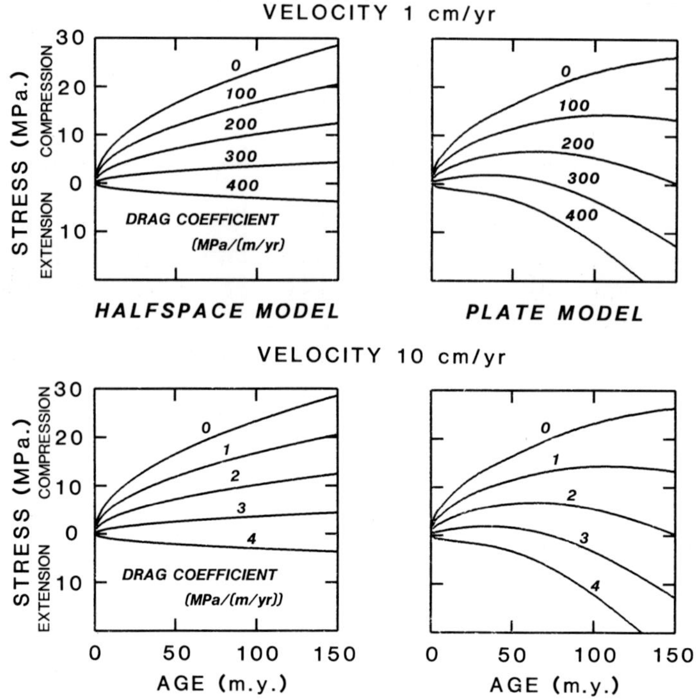

Figure 5. Intraplate stress as a function of lithospheric age and assumed basal drag coefficient for slow moving (top) and fast moving (bottom) plates. Figures on the left assume a half-space thermal model (McKenzie 1969*b*); figures on the right assume a plate model (Parsons & Sclater 1977). The observation of compressional stresses in oceanic lithosphere places an upper bound on the drag coefficient (Wiens & Stein 1985).

need not be a criticism of a model. The generalization that stress in old lithosphere is compressive can constrain the relative magnitudes of the ridge push and basal drag forces. Although the magnitude of ridge push is predicted from models of the thermal evolution of the lithosphere, the magnitude of the basal drag is unknown. An estimate of the drag is useful both for plate driving-force models, and for study of the viscosity of the asthenosphere below the plates.

A simple approach to this issue is to use a two-dimensional model (Wiens & Stein 1985). Assuming the plate is in equilibrium, the stress is determined by the distributed ridge-push body force and boundary conditions representing the effect of material below the plate and at the ridge. Stress at the 'end' boundary is a result rather than a condition; deviatoric compression in old lithosphere indicates that the net force due to the adjacent plate (which in the case of subduction includes the thrust fault contact) is resistive. Initially, the ridge axis is assumed to be rheologically weak, and hence stress free, as discussed later.

The magnitude of the intraplate stress in this model reflects the difference between the effects of ridge push and basal drag. Figure 5 shows intraplate stress as a function of age and drag coefficient, for a half-space model of slow and fast moving plates. If the drag coefficient is zero, the deviatoric stress from ridge push alone is compressive (negative) and varies as the square root of age, because ridge push increases linearly while lithospheric thickness increases as the square root. For larger drag values, the stress follows square-root curves corresponding to less compression, until the lithosphere is in extension for all ages. To satisfy the constraint that old lithosphere is in compression the drag coefficient for the rapidly moving plate must be less than about 3.5 MPa (m $a^{-1})^{-1}$, corresponding to a basal stress less than 0.35 MPa (3.5 bar). Comparable values were found by Hager & O'Connell (1981) and Fleitout & Froidevaux (1983). In contrast, a plate moving a factor of ten slower can remain in compression with a drag coefficient 100 times greater, corresponding to a basal stress ten times greater.

Similar results arise if the lithosphere is modelled as a cooling plate. In this model, because the lithosphere approaches constant thickness with age, the ridge-push force also approaches a constant. As a result, the stress need not be monotonic with age. For no drag, deviatoric stress becomes more compressive with age. For larger values of the drag coefficient the stress reaches a compressional maximum, becomes less compressive, and eventually becomes extensional. For the plate models, the maximum drag coefficient allowing compression in old lithosphere is about half of the value for the half-space models.

Such a model can also yield constraints on mantle viscosity structure (Wiens & Stein 1985) under a simple assumption that the drag on the base of a plate is due to motion over the viscous mantle. For a simple two-dimensional geometry the mass flux due to the moving plate is balanced by a return flow at depth. A drag coefficient consistent with the focal mechanisms would be expected if the plate is underlain by a thin low-viscosity layer, which decouples the plates from the higher-viscosity material below, as proposed by Richter & McKenzie (1978) and Hager & O'Connell (1979).

These two-dimensional models are simple representations of what might occur for an average oceanic plate. None the less, they demonstrate how the observation of intraplate thrust earthquakes in old lithosphere, and the consequent assumption of deviatoric compression, can be used to draw interesting inferences about driving forces. More detailed results can be derived from models in which the forces vary laterally, such that the direction of the compressive stress can be modelled.

4. Young lithosphere

The focal mechanisms for intraplate earthquakes in lithosphere younger than about 35 Ma (figure 3) show both compression and extension (Wiens & Stein 1984; Bergman & Solomon 1984). The simplest interpretation of the extensional events would be that they reflect a broad zone of extension about the extensional mid-ocean ridges. If so, the direction of extension should be normal to the spreading centres. Surprisingly, this does not appear to be the case.

Figure 6 shows that the direction of the horizontal component of the T (tensional) axes for normal faulting events do not indicate extension in the spreading direction. The T axes are oriented at large angles to the spreading direction, and may show a weak preferred orientation approximately perpendicular to the spreading direction. However, since the T axis data-set is strongly dominated by data from the Indian Ocean, it may largely reflect local processes, discussed later. The P axes of near-ridge thrust faulting events show a weak preferred orientation in the direction of spreading.

The absence of ridge-normal extensional earthquakes suggests that the ridge is rheologically weak, as assumed in the previous section. Figure 7 shows the vertically averaged stress in the spreading direction for the model used earlier (§3), with

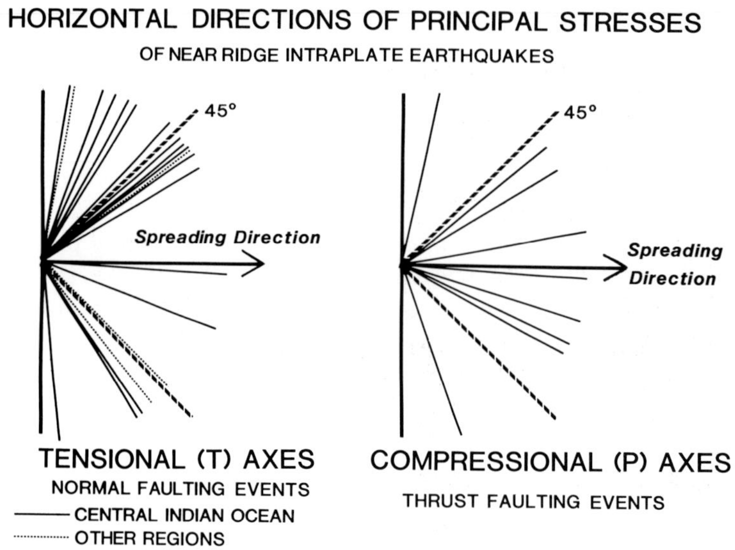

Figure 6. Histograms of the angle between the horizontal component of principal stress and the spreading direction for normal faulting (left) and thrust faulting (right) intraplate earthquakes in young oceanic lithosphere. Tensional axis of normal faulting events, most of which are located in the Central Indian Ocean, generally show extension oblique to the spreading direction, while compressional axis of thrust faulting events show a weak preferred orientation in the spreading direction (Wiens & Stein 1984).

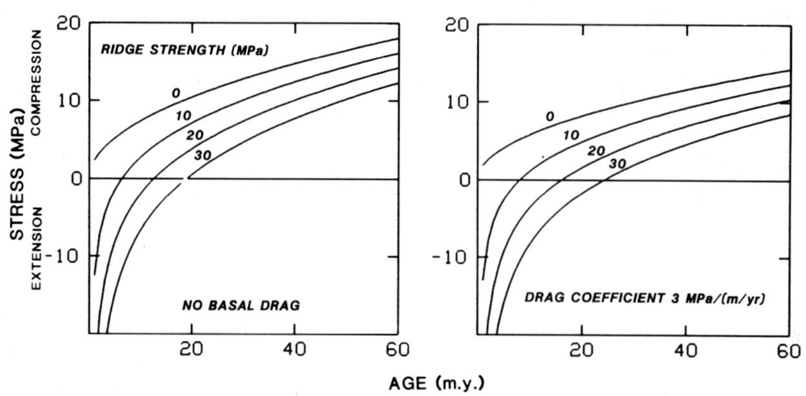

Figure 7. Vertically averaged stress in the spreading direction as a function of lithospheric age computed for a plate velocity of 5 cm a^{-1} and several values of ridge strength and basal drag. The location of the transition from ridge-normal extension to compression is very sensitive to the strength of the ridge (Wiens & Stein 1984).

different boundary conditions at the ridge. The stress field in young lithosphere, especially the location of the transition from compressive to extensional stresses, is sensitive to the strength of the ridge. Ridge axes capable of supporting greater tensional stresses produce wider zones of intraplate extension. Since this ridge-normal extension zone is not observed, the spreading-related extension seems to be restricted to a narrow (less than 100 km) zone about a weak ridge. This constraint is useful for intraplate stress and driving force modelling.

What then gives rise to the ridge-parallel extension? One possibility is thermoelastic stresses (Turcotte & Oxburgh 1973; Parmentier & Haxby 1986;

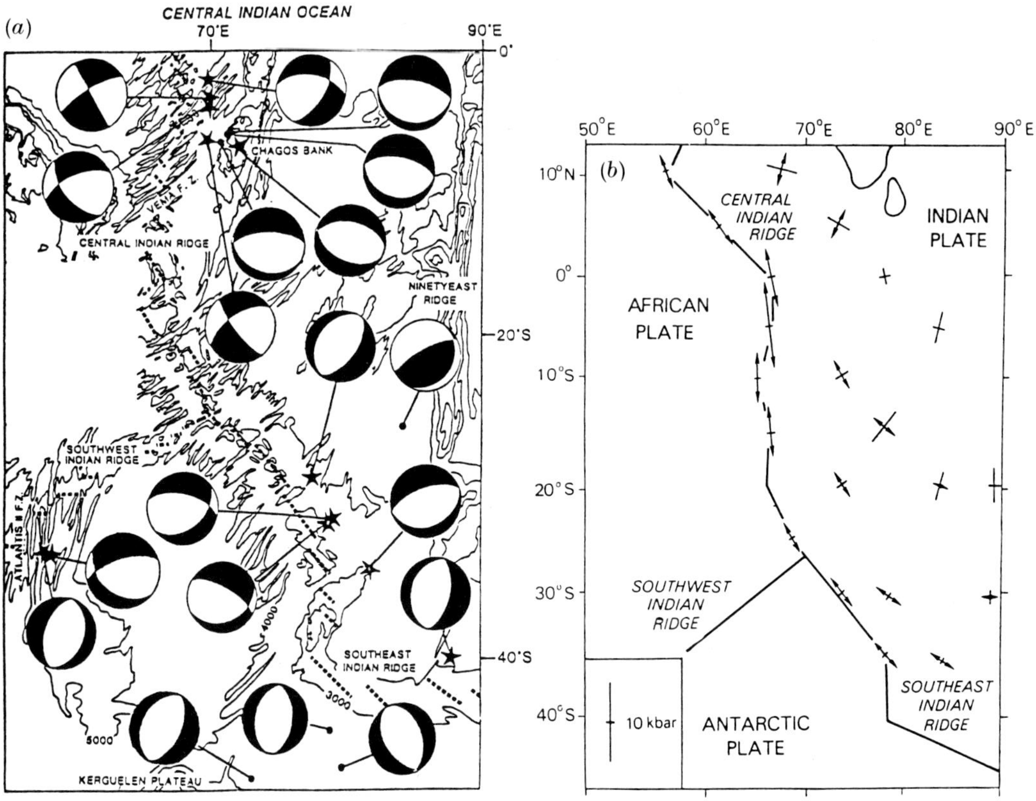

Figure 8. (*a*) Near-ridge intraplate seismicity of the Central Indian Ocean. Most events indicate a regionally consistent extensional stress field oriented parallel to the ridge axis. Dashed lines denote ridge axes. Bathymetry is shown in 1000 m intervals (Wiens & Stein 1984). (*b*) Stresses predicted by model of Cloetingh & Wortel (1986), rescaled for an elastic plate thickness corresponding with the 600 °C isotherm. Symbols (←|→) and (–|–) denote tension and compression, respectively. The predicted stresses are in good agreement with the mechanisms (Stein *et al.* 1987).

Sandwell 1986). An attractive feature of this explanation is that the thermoelastic stresses might decrease with age (Bratt *et al.* 1985), consistent with the decrease in seismicity (figure 2). A difficulty, however, is that the ridge-parallel extensional earthquakes are concentrated in the Central Indian Ocean (figure 8*a*), whereas there is no obvious reason to expect thermoelastic stresses to behave this way.

A more satisfying alternative is that the Indian Ocean extensional zone is a local tectonic situation. Cloetingh & Wortel (1985, 1986) have presented a stress model for a single Indo-Australia plate. The predicted stresses vary dramatically with position, and show a pattern consistent (figure 8*b*) with the focal mechanisms. Moreover, the stress averaged over the mechanically strong lithosphere decreases with age in a manner similar to the normal faulting seismicity (figure 9). It thus appears that the extensional zone can be modelled as a consequence of plate-driving forces. Interestingly, it appears that this situation modelled assuming a single Indo-Australia plate corresponds to the recently recognized diffuse plate-boundary zone between separate Indian and Australian plates (Wiens *et al.* 1985, 1986; Gordon *et al.* 1990). It thus appears that although thermoelastic stresses may provide a low level 'background' in all plates, the Indian Ocean extensional zone is a local phenomenon.

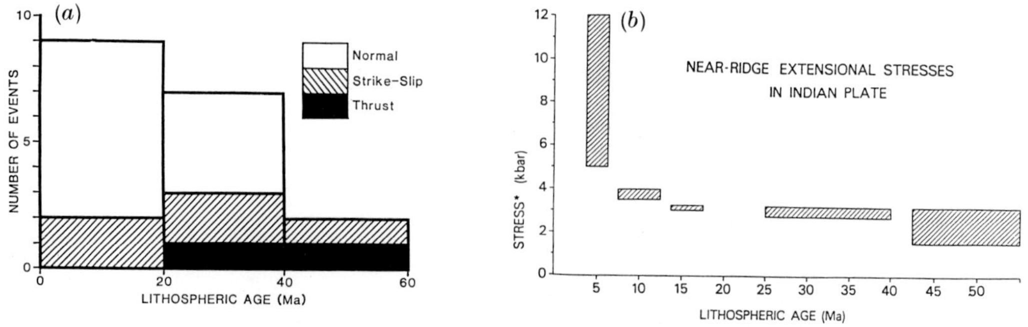

Figure 9. (*a*) Types of faulting found in intraplate earthquakes in the Central Indian Ocean, on the Indian Plate south of 0° and west of 90° E, as a function of lithospheric age. Extensional mechanisms predominate in young lithosphere. (*b*) Average stresses as a function of age from figure 8*a*. The stress decreases in a fashion similar to the number of normal fault earthquakes (Stein *et al.* 1987).

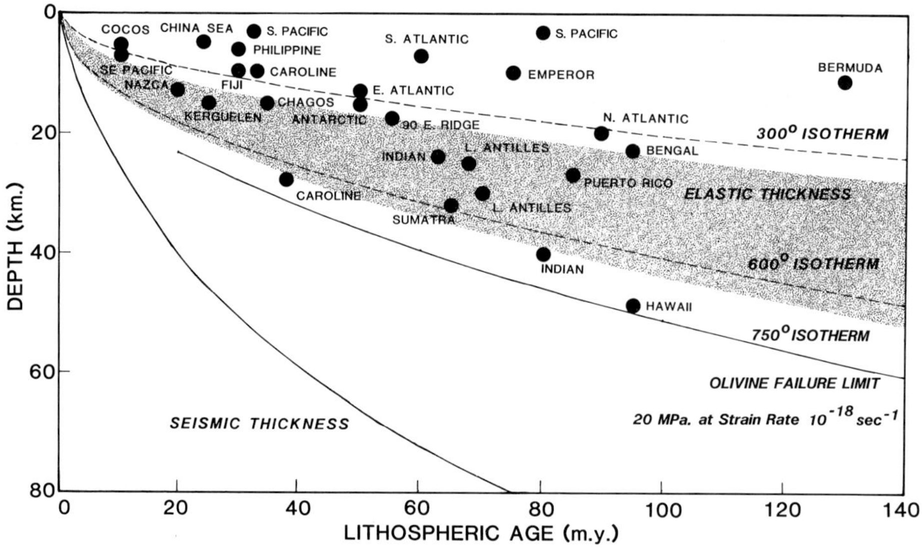

Figure 10. Oceanic earthquake depths against thermal structure of the lithosphere. Isotherms (in degrees centigrade) are for a plate cooling model (Parsons & Sclater 1977). Stippled region denotes estimates of the flexural elastic thickness (Watts *et al.* 1980). The seismic thickness is from surface-wave dispersion data (Leeds *et al.* 1974). The failure limit is the lower limit at which 20 MPa (200 bar) deviatoric stress can be sustained, calculated for a dry olivine rheology (Goetze & Evans 1979) and a strain rate of 10^{-18} s^{-1} (Wiens & Stein 1983).

5. Depths of intraplate earthquakes

The second major data-set derivable from observations of intraplate earthquakes, which has implications for lithospheric stress, is the depths of the earthquakes. In the past decade, waveform modelling techniques have made it possible to determine the focal depths of oceanic intraplate earthquakes to a precision of a few kilometres (Stein & Weins 1986). Figure 10 shows that the maximum depth of seismicity increase with ages. Comparison with the predicted thermal structure of the oceanic lithosphere for a plate cooling model suggests that the maximum depth of seismicity corresponds to a temperature of *ca.* 750 °C (Wiens & Stein 1983; Chen & Molnar

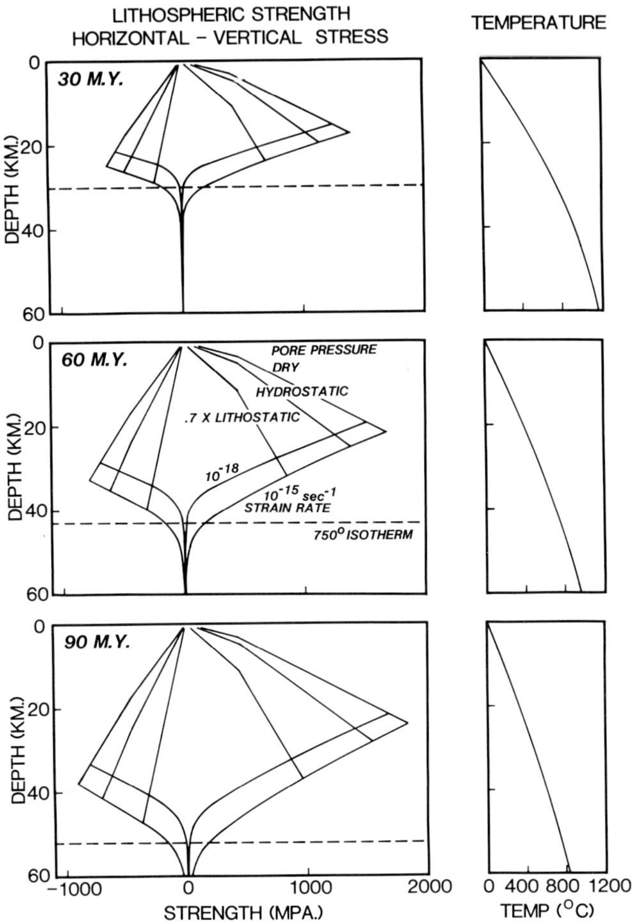

Figure 11. Strength envelopes showing maximum stress difference as a function of depth calculated using temperatures in degrees centigrade from the Parsons & Sclater (1977) plate cooling model. At shallow depth strength is controlled by brittle fracture (Byerlee 1968), at greater depth ductile flow (Goetze & Evans 1979) predicts rapid weakening. The 750 °C isotherm (dashed line) is the approximate lower bound for oceanic intraplate seismicity (Wiens & Stein 1983).

1983). This depth approximately equals the elastic thickness determined from studies of lithospheric flexure due to applied loads, but is significantly less than the depth to the low-velocity zone ('seismic thickness') found from surface-wave dispersion. As the elastic thickness measures the response of the lithosphere to long-term loads, whereas the seismic thickness measures its response to short-term loads, it appears that intraplate earthquakes occur primarily where the lithosphere can sustain long-term loads.

The focal depths suggest that above a temperature of *ca.* 750 °C the lithosphere is too weak to support the stresses needed for seismic failure. This effect can be modelled using results from laboratory rock mechanics studies (Kirby 1977, 1980; Goetze & Evans 1979; Brace & Kohlstedt 1980) to predict the variation of strength (the maximum difference between principal stresses that rock can support) with depth in the lithosphere. At low temperatures, corresponding to shallow depths, failure occurs by brittle fracture and strength increases with depth. At higher

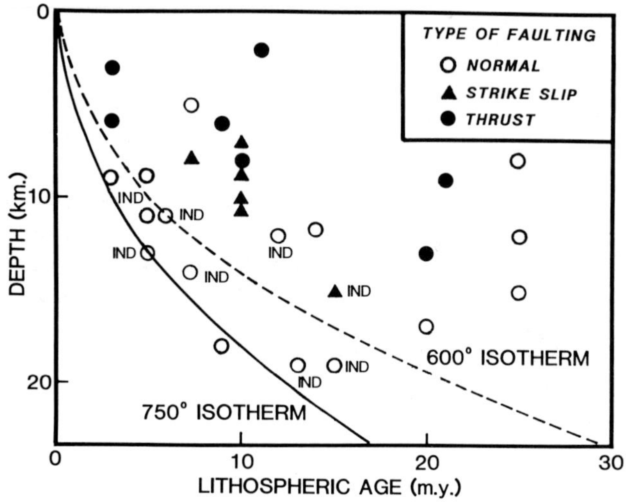

Figure 12. Depths of intraplate earthquakes in young lithosphere as a function of lithospheric age. Isotherms are from the Parsons & Sclater (1977) plate cooling model. Normal faulting events seem to be located at greater depths and temperatures than thrust and strike slip events. Most comparatively deep normal faulting occurs in the Indian Plate (events marked 'IND') (Wiens & Stein 1984).

temperatures, where rocks deform by ductile flow, strength decreases exponentially with temperature. Hence as the lithosphere cools, the portion expected to have strength adequate to support seismicity thickens (figure 11).

Comparison of the predicted strengths to the depth observations requires a specific rheology. It also requires the assumption of a strain rate, because strength in the ductile region increases with strain rate, and a pore pressure distribution, because in the brittle region strength decreases with pore pressure. None the less, the basic idea is robust. An interesting feature of this analysis is that, for the rheology and thermal structure assumed, earthquakes occur in both the brittle and ductile regions. If the parameters assumed are appropriate, it appears that the requirement for seismicity is adequate strength, rather than a brittle rheology. Another interesting implication is that normal faulting may occur deeper than thrust faulting, because the lithosphere is stronger to a greater depth in deviatoric extension than in compression. The focal depth data provide a suggestion of this phenomenon (figure 12), although the generally greater depths of the normal fault events could reflect another cause.

This type of strength envelope analysis has become a standard tool in investigating lithospheric stresses. Observations of either the maximum depth of earthquakes or the thickness of the mechanically strong portion of the lithosphere (Bodine *et al.* 1981; McNutt & Menard 1982) are combined with an assumed temperature structure to constrain stresses. This has been done to study regions of intraplate deformation (Stein & Weissel 1990), oceanic transform faulting (Engeln *et al.* 1986; Bergman & Solomon 1988; Chen 1988), and possible midplate reheating (Stein & Abbott 1991). A similar technique is also used extensively in studies of continental geology (Sibson 1982; Meissner & Strehlau 1982; Smith & Bruhn 1984; Kusznir & Karner 1985; Shudofsky *et al.* 1987).

The technique has also been applied recently to the question raised earlier (§4) about the strength of the mid-ocean ridge. Mid-ocean ridge normal fault earthquakes

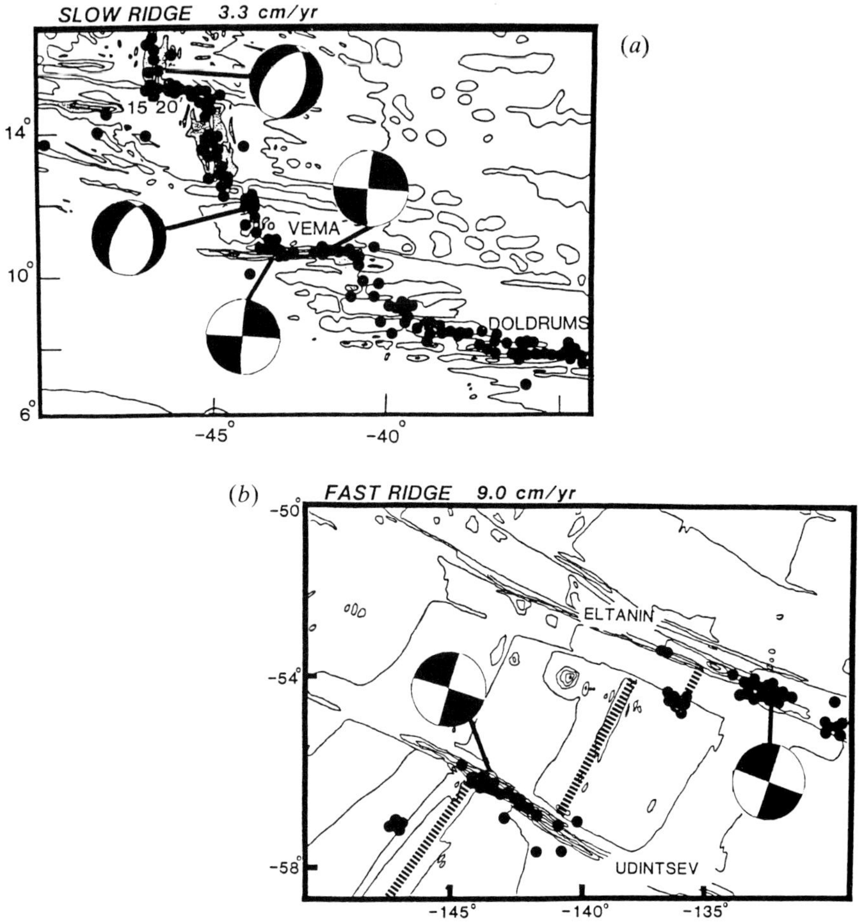

Figure 13. Comparison of earthquake mechanisms on slow and fast spreading ridges. The slow spreading Mid-Atlantic ridge (*a*) has earthquakes with normal faulting on the ridge axis, and strike–slip earthquakes on the transforms. In contrast, the fast spreading East Pacific Rise (*b*) shows no seismicity along the ridge segments, although strike–slip earthquakes occur on the transforms (Stein & Woods 1989.)

are restricted to slow spreading ridges (half rate less than 3 cm a^{-1}) (figure 13), where the axis is characterized by a median valley (Sleep & Rosendahl 1979). The maximum depth of the earthquakes decreases with spreading rate (Huang & Solomon 1988). This trend is plausible, since thermal models (Sleep 1975) predict that the ridge axis will be cooler to greater depth for slower spreading. The earthquakes extend, however, to depths below the predicted 750 °C isotherm. As a result, the ridge axis should have insufficient strength to permit this seismicity. Lin & Parmentier (1989) thus propose that the ridge is cooled sufficiently by hydrothermal circulation that seismicity can occur to these depths.

A final point worth noting is that such analyses use the earthquake depths only to infer that strength at that depth exceeds a minimum required value. A recent approach (Govers *et al.* 1991) explicitly estimates the magnitude of the tectonic stress at depth by assuming that the earthquake's occurrence indicates that stress has reached the predicted strength envelope.

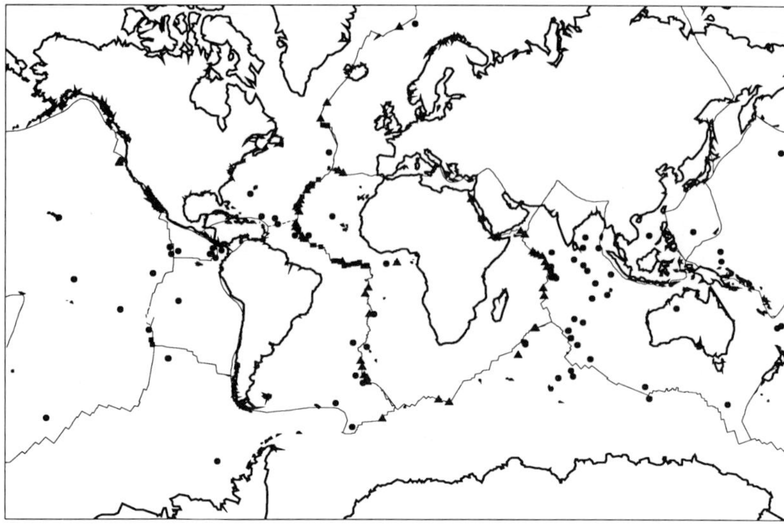

Figure 14. Locations of earthquakes used in figures 15 and 16. Intraplate events are shown by circles; transform events by squares; and mid-ocean ridge events by triangles. These events are studied by various investigators: Wiens & Stein (1984), Bergman *et al.* (1985), Bergman & Solomon (1984, 1985), and Bergman (1986) for the intraplate earthquakes; Engeln (1985), Engeln *et al.* (1986), Goff *et al.* (1987), and Bergman & Solomon (1988) for transform events; Huang *et al.* (1986), Huang & Solomon (1987, 1988) and Jemsek *et al.* (1986) for mid-ocean ridge earthquakes.

6. Source parameters

In addition to earthquake mechanisms and focal depths, seismological observations may be able to provide useful information on the magnitudes of stress via parameters characterizing the earthquake rupture. Seismic source theory (reviewed by Geller (1976) and Kanamori (1980)) provides useful insight. The seismic radiation from an earthquake depends on parameters including the seismic moment

$$M_0 = \mu \bar{D} S, \tag{6.1}$$

where μ is the rigidity and $\bar{D}$ is the average slip (or dislocation) on the fault, whose area is S. For continental or large subduction zone earthquakes, the fault area can often be inferred from aftershocks. In contrast, for oceanic intraplate events, the fault area is usually inferred from the duration of the pulse τ generated at the source. This duration depends on how long the rupture takes to propagate across the fault area, which depends on the rupture velocity, a quantity assumed to be proportional to the shear velocity β

$$\tau = cS^{\frac{1}{2}}/\beta. \tag{6.2}$$

Since the strain change associated with the earthquake is proportional to the ratio of the slip to the square root of the fault area, the corresponding stress drop is

$$\Delta\sigma = C\mu\bar{D}/S^{\frac{1}{2}} = CM_0/S^{\frac{3}{2}} \approx CM_0/(\beta\tau)^3, \tag{6.3}$$

where C depends on the fault geometry and other parameters. Hence for a given seismic moment, longer source time functions correspond to lower stress drops. Although there are further complexities, due to the finite rise time required for the dislocation to reach its full value at any point, this general concept can be used to estimate relative stress drops (Chung & Kanamori 1980). Note that this method

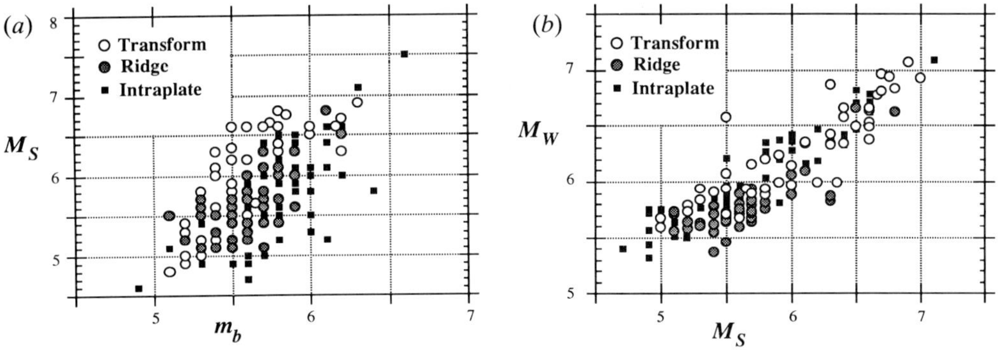

Figure 15. Comparison of body and surface wave magnitudes (*a*) and surface wave and moment magnitudes (*b*) for oceanic ridge, transform and intraplate earthquakes. Both comparisons show that the transform events release more energy at longer periods than ridge events, whereas intraplate events overlap both.

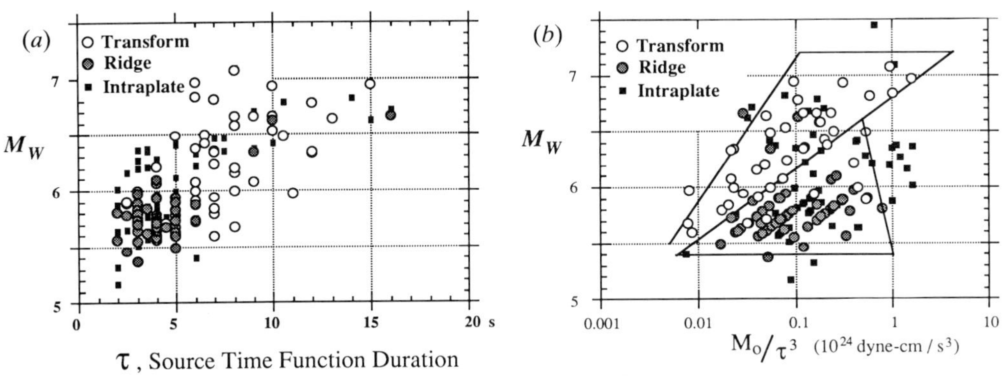

Figure 16. Comparison of time function duration τ and M_0/τ^3 as a function of moment magnitude M_w. For a given M_w, transform earthquakes have longer time functions and hence smaller M_0/τ^3 than ridge events. This effect may indicate that ridge events may have higher stress drops than transform events.

examines only stress drops, rather than the ambient tectonic stress. The relation between the two depends on assumption about the rupture process.

To date, studies of oceanic earthquakes have used a related concept, comparison of the magnitudes determined at different periods, to estimate how energy release varies with period. For this purpose, the body wave magnitude determined at a period of 1 s (m_b), the surface wave magnitude determined at a period of 20 s (M_s), and the moment magnitude determined from the longest period data possible (M_w) (Kanamori 1977) can be compared. Kanamori & Stewart (1976) and Okal & Stewart (1982) suggest that some oceanic transform earthquakes have large M_s relative to m_b, suggesting a slow energy release, perhaps qualitatively indicative of low stress.

We have begun (Pelayo & Stein 1991) to apply this idea to a data-set of oceanic ridge, transform and intraplate earthquakes (figure 14). This data-set has the advantage that all the earthquakes occur in oceanic lithosphere, so we compare events in material of presumably similar mechanical properties, rather than compare them with global data-sets that contain many continental earthquakes. The preliminary results show several interesting patterns (figures 15, 16). For a given m_b, transform events generally have higher M_s than ridge events, suggesting that more energy is released at longer periods. Intraplate events overlap both the ridge and transform events. A similar pattern emerges for M_s and M_w. A consistent pattern can be seen in another way, because for a given M_w, transform earthquakes have longer time functions and hence smaller M_0/τ^3 than ridge events. Thus with respect to the population of intraplate events, ridge earthquakes are generally 'fast' whereas transform earthquakes are generally 'slow'. A variety of factors may contribute to these variations. Ridge events may have higher stress drops than transform events, perhaps due to higher tectonic stress. Alternatively, these differences may reflect different fault geometries or rupture properties. We hope to determine which of the possibilities is most likely.

This research was supported by NSF grant EAR 8618038 and NASA Crustal Dynamics Grant NAG5-885.

References

Aki, K. & Richards, P. G. 1980 *Quantitative seismology: theory and methods*. San Francisco: W. H. Freeman.

Bergman, E. A. 1986 Intraplate earthquakes and the state of stress in oceanic lithosphere. *Tectonophys*. **132**, 1–35.

Bergman, E. A. & Solomon, S. C. 1984 Source mechanisms of earthquakes near mid-ocean ridges from body waveform inversion: implications for the early evolution of oceanic lithosphere. *J. geophys. Res.* **89**, 11415–11441.

Bergman, E. A. & Solomon, S. C. 1985 Earthquake source mechanisms from body waveform inversion and intraplate tectonics in the northern Indian Ocean. *Phys. Earth planet. Interiors* **40**, 1–23.

Bergman, E. A. & Solomon, S. C. 1988 Transform fault earthquakes in the North Atlantic: source mechanisms and depth of faulting. *J. geophys. Res.* **93**, 9027–9057.

Bergman, E. A., Nabelek, J. L. & Solomon, S. C. 1985 Correction to 'An extensive region of off ridge normal-faulting earthquakes in the southern Indian Ocean'. *J. geophys. Res.* **90**, 2066.

Bodine, J. H., Steckler, M. S. & Watts, A. B. 1981 Observations of flexure and the rheology of the oceanic lithosphere. *J. geophys. Res.* **86**, 3695–3707.

Brace, W. F. & Kohlstedt, D. L. 1980 Limits on lithospheric stress imposed by laboratory experiments. *J. geophys. Res.* **85**, 6248–6252.

Byerlee, J. D. 1968 Brittle–ductile transition in rocks. *J. geophys. Res.* **73**, 4741–4750.

Bratt, S. R., Bergman, E. A. & Solomon, S. C. 1985 Thermoelastic stress: how important as a cause of earthquakes in young oceanic lithosphere? *J. geophys. Res.* **90**, 10249–10260.

Chen, W.-P. & Molnar, P. 1983 Focal depths of intracontinental and intraplate earthquakes and their implications for the thermal and mechanical properties of the lithosphere. *J. geophys. Res.* **88**, 4183–4214.

Chen, Y. 1988 Thermal model of oceanic transform faults. *J. geophys. Res.* **93**, 8839–8851.

Chung, W.-Y. & Kanamori, H. 1980 Variation of seismic source parameters and stress drops within a descending slab and its implications in plate mechanics. *Phys. Earth planet. Interiors* **23**, 134–159.

Cloetingh, S. & Wortel, R. 1985 Regional stress field of the Indian plate. *Geophys. Res. Lett.* **12**, 77–80.

Cloetingh, S. & Wortel, R 1986 Stress in the Indo-Australian plate. *Tectonophys*. **132**, 49–67.

Engeln, J. F. 1985 Seismological studies of the tectonics of divergent plate boundaries. Ph.D. thesis, Northwestern University, Evanston, Illinois.

Engeln, J. F., Wiens, D. A. & Stein, S. 1986 Mechanisms and depths of Atlantic transform earthquakes. *J. geophys. Res.* **91**, 548–577.

Fleitout, L. & Froidevaux, C. 1983 Tectonic stresses in the lithosphere. *Tectonics* **2**, 315–324.

Forsyth, D. W. 1973 Compressive stress between two mid-ocean ridges. *Nature, Lond.* **243**, 78–79.

Forsyth, D. W. & Uyeda, S. 1975 On the relative importance of the driving forces of plate motion. *Geophys. Jl R. astr. Soc.* **43**, 163–200.

Geller, R. J. 1976 Scaling relations for earthquake source parameters and magnitudes. *Bull. seismol. Soc. Am.* **66**, 1501–1523.

Goetze, C. & Evans, B. 1979 Stress and temperature in the bending lithosphere as constrained by experimental rock mechanics. *Geophys. Jl R. astr. Soc.* **59**, 463–478.

Goff, J. A., Bergman, E. A. & Solomon, S. C. 1987 Earthquake source mechanisms and transform fault tectonics in the Gulf of California. *J. geophys. Res.* **92**, 10485–10510.

Gordon, R. G., DeMets, C. & Argus, D. F. 1990 Kinematic constraints on distributed lithospheric deformation in the equatorial Indian Ocean from present motion between the Australian and Indian plates. *Tectonics* **9**, 409–422.

Govers, R., Wortel, R., Cloetingh, S. & Stein, C. 1991 Stress magnitude estimates from earthquakes in oceanic plate interiors. *J. geophys. Res.* (Submitted.)

Hager, B. H. & O'Connell, R. J. 1979 Kinematic models of large-scale flow in the Earth's mantle. *J. geophys. Res.* **84**, 1031–1048.

Hager, B. H. & O'Connell, R. J. 1981 A simple global model of plate dynamics and mantle convection. *J. geophys. Res.* **86**, 4843–4867.

Handin, J. 1969 On the Coulomb–Mohr failure criterion. *J. geophys. Res.* **74**, 5343–5348.

Harper, J. F. 1975 On the driving forces of plate tectonics. *Geophys. Jl R. astr. Soc.* **40**, 465–474.

Huang, P. Y. & Solomon, S. C. 1987 Centroid depths and mechanisms of ridge-axis earthquakes in the Indian Ocean, Gulf of Aden, and Red Sea. *J. geophys. Res.* **92**, 1361–1382.

Huang, P. Y. & Solomon, S. C. 1988 Centroid depths of mid-ocean ridge earthquakes: dependence on spreading rate. *J. geophys. Res.* **93**, 13445–13477.

Huang, P. Y., Solomon, S. C., Bergman, E. A. & Nabelek, J. L. 1986 Focal depths and mechanisms of Mid-Atlantic Ridge earthquakes from body waveform inversion. *J. geophys. Res.* **91**, 579–598.

Jaeger, J. C. & Cook, N. G. W. 1976 *Fundamentals of rock mechanics*. London: Chapman & Hall.

Jemsek, J. P., Bergman, E. A., Nabelek, J. L. & Solomon, S. C. 1986 Focal depths and mechanisms of earthquakes on the Arctic mid-ocean ridge system. *J. geophys. Res.* **91**, 13993–14005.

Kanamori, H. 1977 Seismic and aseismic slip along subduction zones and their tectonic implications. In *Island arcs, deep-sea trenches and back-arc basins, Maurice Ewing Ser.*, vol. 1 (ed. M. Talwani & W. C. Pitman III), pp. 163–174. Washington, D.C.: AGU.

Kanamori, H. 1980 State of stress in the earth's lithosphere. In *Physics of the Earth's interior, Proc. Int. Sch. Phys. 'Enrico Fermi'* (ed. A. M. Dziewonski & E. Boschi), pp. 531–552. Amsterdam: North Holland.

Kanamori, H. & Stewart, G. S. 1976 Mode of the strain release along the Gibbs Fracture Zone, Mid-Atlantic Ridge. *Phys. Earth planet. Interiors* **11**, 312–332.

Kirby, S. H. 1977 State of stress in the lithosphere: Inferences from the flow laws of olivine. *Pure appl. Geophys.* **115**, 245–258.

Kirby, S. H. 1980 Tectonic stresses in the lithosphere: Constraints provided by the experimental deformation of rocks. *J. geophys. Res.* **85**, 6353–6363.

Kusznir, N. & Karner, G. 1985 Dependence of the flexural rigidity of the continental lithosphere on rheology and temperature. *Nature, Lond.* **316**, 138–142.

Leeds, A. R., Knopoff, L. & Kausel, E. G. 1974 Variations of upper mantle structure under the Pacific Ocean. *Science, Wash.* **186**, 141–143.

Lin, J. & Parmentier, E. M. 1989 Mechanisms of lithospheric extension at mid-ocean ridges. *Geophys. J.* **96**, 1–22.

Lister, C. R. B. 1975 Gravitational drive on oceanic plates caused by thermal contraction. *Nature, Lond.* **257**, 663–665.

McKenzie, D. P. 1969*a* The relation between fault plane solutions for earthquakes and the directions of the principal stresses. *Bull. seismol. Soc. Am.* **59**, 591–601.

McKenzie, D. P. 1969*b* Speculations on the consequences and causes of plate motions. *Geophys. Jl R. astr. Soc.* **18**, 1–32.

McNutt, M. K. & Menard, H. W. 1982 Constraints on yield strength in the oceanic lithosphere derived from observations of flexure. *Geophys. Jl R. astr. Soc.* **71**, 363–394.

Meissner, R. & Strehlau, J. 1982 Limits of stresses in continental crusts and their relation to the depth-frequency distribution of shallow earthquakes. *Tectonics* **1**, 73–89.

Mendiguren, J. A. 1971 Focal mechanism of a shock in the middle of the Nazca plate. *J. geophys. Res.* **76**, 3861–3879.

Mendiguren, J. A. & Richter, F. M. 1978 On the origin of compressional intraplate stresses in South America. *Phys. Earth planet. Interiors* **16**, 318–326.

Okal, E. A. & Stewart, L. M. 1982 Slow earthquakes along oceanic fracture zones: evidence for asthenospheric flow away from hot spots. *Earth planet. Sci. Lett.* **57**, 75–87.

Parmentier, E. M. & Haxby, W. F. 1986 Thermal stresses in the oceanic lithosphere: evidence from geoid anomalies at fracture zones. *J. geophys. Res.* **91**, 7193–7204.

Parsons, B. 1982 Causes and consequences of the relation between area and age of the ocean floor. *J. geophys. Res.* **87**, 289–302.

Parsons, B. & Richter, F. M. 1980 A relation between the driving force and geoid anomaly associated with mid-ocean ridges. *Earth planet. Sci. Lett.* **51**, 445–450.

Parsons, B. & Sclater, J. G. 1977 An analysis of the variation of ocean floor bathymetry and heat flow with age. *J. geophys. Res.* **82**, 803–827.

Pelayo, A. M. & Stein, S. 1991 Variation in source parameters for oceanic ridge, transform and intraplate earthquakes (abstract). *Eos Trans. Am. Geophys. Union* (*Spring Meeting Issue*), p. 202.

Richardson, R. M. & Cox, B. L. 1984 Evolution of oceanic lithosphere: a driving force study of the Nazca plate. *J. geophys. Res.* **89**, 10043–10052.

Richardson, R. M., Solomon, S. C. & Sleep, N. H. 1976 Intraplate stress as an indicator of plate tectonic driving forces. *J. geophys. Res.* **81**, 1847–1856.

Richardson, R. M., Solomon, S. C. & Sleep, N. H. 1979 Tectonic stress in the plates. *Rev. Geophys. Space Phys.* **17**, 981–1019.

Richter, F. M. 1977 On the driving mechanism of plate tectonics. *Tectonophys.* **38**, 61–88.

Richter, F. & McKenzie, D. 1978 Simple plate models of mantle convection. *J. Geophys.* **44**, 441–471.

Sandwell, D. T. 1986 Thermal stress and the spacings of transform faults. *J. geophys. Res.* **91**, 6405–6418.

Shudofsky, G., Cloetingh, S., Stein, S. & Wortel, R. 1987 Unusually deep earthquakes in East Africa: constraints on the thermo-mechanical structure of a continental rift system. *Geophys. Res. Lett.* **14**, 741–744.

Sibson, R. H. 1982 Fault zone models, heat flow, and the depth distribution of earthquakes in the continental crust of the United States. *Bull. seismol. Soc. Am.* **72**, 151–163.

Sleep, N. H. 1975 Formation of oceanic crust: some thermal constraints. *J. geophys. Res.* **80**, 4037–4042.

Sleep, N. H. & Rosendahl, B. R. 1979 Topography and tectonics of mid-oceanic ridge axes. *J. geophys. Res.* **84**, 6831–6839.

Smith, R. B. & Bruhn, R. L. 1984 Intraplate extensional tectonics on the eastern Basin-Range: inferences of structural style from seismic reflection data, regional tectonics, and thermal-mechanical models of brittle–ductile deformation. *J. geophys. Res.* **89**, 5733–5762.

Solomon, S. C. & Sleep, N. H. 1974 Some simple physical models for absolute plate motions. *J. geophys. Res.* **79**, 2557–2567.

Solomon, S. C., Sleep, N. H. & Richardson, R. M. 1975 On the forces driving plate tectonics: inferences from absolute plate velocities and intraplate stress. *Geophys. Jl R. astr. Soc.* **42**, 769–801.

Stein, S. 1979 Intraplate seismicity on bathymetric features: the 1968 Emperor Trough earthquake. *J. geophys. Res.* **84**, 4763–4768.

Stein, C. A. & Abbott, D. H. 1991 Heat flow constraints on the South Pacific superswell. *J. geophys. Res.* (In the press.)

Stein, C. A. & Weissel, J. K. 1990 Constraints on the Central Indian Basin thermal structure from heat flow, seismicity and bathymetry. *Tectonophys.* **176**, 315–332.

Stein, S. & Wiens, D. 1986 Depth determination for shallow teleseismic earthquakes: methods and results. *Rev. Geophys. Space Phys.* **24**, 806–832.

Stein, S. & Woods, D. F. 1989 Seismicity: midocean ridge. In *The encyclopedia of solid Earth geophysics* (ed. D. E. James), pp. 1050–1054. New York: Van Nostrand Reinhold.

Stein, S., Cloetingh, S., Sleep, N. & Wortel, R. 1989 Passive margin earthquakes, stresses, and rheology. In *Earthquakes at North-Atlantic passive margins: neotectonics and postglacial rebound* (ed. S. Gregerson & P. Basham), pp. 231–260. Reidel.

Stein, S., Cloetingh, S., Wiens, D. & Wortel, R. 1987 Why does near-ridge extensional seismicity occur primarily in the Indian Ocean? *Earth planet. Sci. Lett.* **82**, 107–113.

Sykes, L. R. & Sbar, M. L. 1974 Focal mechanism solutions of intraplate earthquakes and stresses in the lithosphere. In *Geodynamics of Iceland and the North Atlantic area* (ed. L. Kristjansson), pp. 207–224. Dordrecht: Reidel.

Turcotte, D. L. & Oxburgh, E. R. 1973 Mid-plate tectonics. *Nature, Lond.* **244**, 337–339.

Watts, A. B., Bodine, J. H. & Steckler, M. S. 1980 Observations of flexure and the state of stress in the oceanic lithosphere. *J. geophys. Res.* **85**, 6369–6376.

Wiens, D. A. & Stein, S. 1983 Age dependence of oceanic seismicity and implications for lithospheric evolution. *J. geophys. Res.* **88**, 6455–6468.

Wiens, D. A. & Stein, S. 1984 Intraplate seismicity and stresses in young oceanic lithosphere. *J. geophys. Res.* **89**, 11442–11464.

Wiens, D. A. & Stein, S. 1985 Implications of oceanic intraplate seismicity for plate stresses, driving forces and rheology. *Tectonophys.* **116**, 143–162.

Wiens, D. A., DeMets, C., Gordon, R. G., Stein, S., Argus, D., Engeln, J. F., Lundgren, P., Quible, D., Stein, C., Weinstein, S. & Woods, D. F. 1985 A diffuse plate boundary model for Indian Ocean tectonics. *Geophys. Res. Lett.* **12**, 429–432.

Wiens, D., Stein, S., DeMets, C., Gordon, R. & Stein, C. 1986 Plate tectonic models for Indian Ocean 'intraplate' deformation. *Tectonophys.* **132**, 37–48.

Wortel, M. J. R. & Cloetingh, S. A. P. L. 1985 Accretion and lateral variations in tectonic structure along the Peru–Chile Trench. *Tectonophys.* **112**, 443–462.

Zoback, M. L., *et al.* 1989 Global patterns of tectonic stress. *Nature, Lond.* **341**, 291–298.

Discussion

R. Wortel (*Utrecht University, The Netherlands*). Is it not valid to use focal mechanisms as a constraint on modelling basal shear stresses only for a cylindrical earth? Is not the generally adopted idea that old oceanic lithosphere is in compression based on a very limited number of data points?

S. Stein. Although the number of focal mechanisms available for old oceanic lithosphere is quite limited, they indicate compression. Whether these sparse points are an adequate sample is unknown. Given the sparse data, we used a simple two-dimensional model to deal with average properties versus age. Certainly a three-dimensional model would be more appropriate.

L. Fleitout (*École Normale Supérieure, Paris, France*). Could thermal stresses be the cause of the compression in the ocean for all ages and the focal mechanisms that Dr Stein sees do not reflect tectonic mechanisms? Thermal stresses should increase with age and the Pacific Plate can support seamounts for tens of millions of years so I do not see why the oceanic lithosphere should relieve its thermal stresses.

S. Stein. I agree that thermal stress may well contribute. Some features of the focal mechanism data, notably ridge-parallel extension in young lithosphere, are suggestive of some thermal stress contribution. A difficulty, however, is that the ridge-parallel extensional earthquakes are concentrated in the Central Indian Ocean whereas there is no obvious reason to expect thermoelastic stresses to behave this way. For the old lithosphere, the fact that the compressivė stresses can be explained without thermal stress does not preclude thermal stress being a contributor.

S. Murrell (*University College, London, U.K.*). Is there any memory in the oceanic lithosphere of the original extension?

S. Stein. The lithosphere preserves evidence of the ridge-normal extension at the axis, via ridge-parallel tectonic fabric which can be used to model the extension. I know of no preserved evidence of the ridge-parallel extension in young lithosphere.

J. Bull (*Southampton University, U.K.*). In the central Indian Ocean Basin there is very clear evidence of the reactivation of pre-existing structures but there is also a problem. On seismic reflection profiles you can see that there are reactivated features going down to 5 or 6 km. So either the faulting has propagated downwards or else there has been nucleation of earthquakes at the brittle–ductile transition and then propagation upwards and reactivation of the pre-existing structures.

S. Stein. I agree that there is often reactivation of pre-existing structures, although we know little about the details.

E. J. W. Jones (*University College, London, U.K.*). I think one has to be very careful about using earthquakes alone to infer the stress régime in the oceans because if high-resolution seismic reflection profiles from the North Atlantic are examined, one finds that in abyssal plains near continental margins there is a great deal of very recent deformation in areas in which there isn't apparently any clear seismicity.

S. Stein. The fact that we see deformation in regions where we haven't observed seismicity may not be a problem, given the very short time interval over which we have instrumental records.

M. Zoback (*Stanford University, U.S.A.*). I am concerned about the big discrepancy between what is normally considered to be the brittle–ductile transition, on the basis of the rheological models, and the maximum depths of the earthquakes. This seems to be almost a factor of two.

S. Stein. The intraplate oceanic earthquakes' focal depths extend well below the brittle–ductile transition predicted from the laboratory rheology. It is possible that the rheology is inappropriate. If this is not the case, it appears that the depth of seismicity is limited not by the brittle–ductile transition, but by the depth to which the lithosphere has adequate strength. In the continents, these two depths are similar, assuming that a quartz rheology is appropriate, so it is hard to tell the difference. The idea of earthquakes below the brittle–ductile transition seems plausible, given that in subduction zones there are earthquakes below the transition.

The sources of lithospheric tectonic stresses

By Luce Fleitout

Laboratoire de Géologie-ENS, 24 rue Lhomond, 75005 Paris, France

The stresses associated with large-scale tectonic deformation have three possible origins: (1) plate-boundary forces counterbalanced by viscous drag beneath the plates; (2) density heterogeneities situated within the plates (say at depths shallower than 200 km); (3) mass heterogeneities in the deep mantle. The first two are shown to be equally important for the understanding of the stress field. No topography (no vertical stress) seems to be associated with lower-mantle mass anomalies. This is most compatible with a two-layer convective mantle where the lower-mantle mass anomalies, mechanically decoupled from the lithosphere, are unable to induce tectonic stresses.

1. Introduction

Large-scale tectonic stresses were first interpreted in terms of plate-boundary forces (ridge push, slab pull, etc.), counterbalanced by the viscous drag beneath the plates. Intralithospheric mass heterogeneities modify the vertical stresses in the lithosphere and are thus an important element of the tectonic stress field (Artyushkov 1973; Fleitout & Froidevaux 1982). Deeper mass heterogeneities in the convective mantle are also potentially important ingredients of the stress field. The purpose of this paper is to discuss the relative importance of these various sources for the stress field.

This paper deals only with 'tectonic stresses', those able to induce large-scale tectonic deformation, i.e. stresses which are not released by a small (less than 1%) plastic deformation, such as flexural or thermal stresses. The voluntarily simplified models developed here best apply to relatively large-scale tectonic deformation and do not take into account the perturbations which can arise in a three-dimensional geometry from laterally variable mechanical properties or boundary conditions.

2. Density anomalies in the lithosphere and tectonic stresses

(a) The 'moment' law

Let us consider a thin mass anomaly Δm, situated in the lithosphere at a depth d. How does it influence the stress field? The exact solutions for a lithosphere made of viscous layers are presented by Fleitout & Froidevaux (1982). Here we present a simple explanation for the results obtained for 'large wavelengths', i.e. in cases where the lateral extent of the mass anomaly dx is large with respect to the thickness of the lithosphere L. Then, the horizontal strain rate ϵ_{xx} is independent of z (the difference between σ_{xx} and σ_{zz} varies as a function of depth so that ϵ_{xx} remains constant). The strain rate is then only a function of the integral

$$\Sigma = \int_0^L (\sigma_{xx} - \sigma_{zz})\,dz. \tag{1}$$

The horizontal equilibrium of a plate section is written as

$$\frac{\partial}{\partial x}\int_0^L \sigma_{xx}\,\mathrm{d}z = -\sigma_{xzL}. \tag{2}$$

Equation (2) implies that over short distances, $\int_0^L \sigma_{xx}\,\mathrm{d}z$ can be considered constant. Variations of the tectonic style between nearby areas only arise from variations of the averaged vertical stress.

How is the vertical stress σ_{zz} affected by the mass anomaly? Let us examine in figure 1 the difference between the vertical stresses over column B, away from the mass anomaly δm and a column containing the mass anomaly (column A).

At the depth L, σ_{zz} is the same for the two columns (similar pressure at a given depth in the fluid sublithospheric mantle). The vertical momentum equation is written as

$$\partial\sigma_{zz}/\partial z + \partial\sigma_{xz}/\partial x = -\Delta\rho\, g, \tag{3}$$

where $\Delta\rho$ is the local density anomaly. (We note that extensional stresses are positive.)

When the depth of the mass anomaly is small compared with its lateral extent $\partial\sigma_{xz}/\partial x$ can be neglected (see Fleitout & Froidevaux (1982) for cases where this condition is not verified). Then, (3) reduces to

$$\partial\sigma_{zz}/\partial z = -\Delta\rho\, g.$$

$\partial\sigma_{zz}/\partial z$ is then the same for the two columns, except for $z = d$:

$$\partial(\sigma_{zz\mathrm{A}} - \sigma_{zz\mathrm{B}})/\partial z = 0 \quad \text{for} \quad z \neq d. \tag{4}$$

$\sigma_{zz\mathrm{A}} - \sigma_{zz\mathrm{B}}$ jumps by $\delta m g$ in $z = d$. $\sigma_{zz\mathrm{A}} - \sigma_{zz\mathrm{B}}$ as a function of depth is presented on figure 1. It follows:

$$\int_0^L (\sigma_{zz\mathrm{A}} - \sigma_{zz\mathrm{B}})\,\mathrm{d}z = \delta m g d. \tag{5}$$

The influence of an intralithospheric mass anomaly on the stress field is therefore proportional to its moment, which is the product of its amplitude and its depth. (This corresponds indeed to the 'moment' of the mass dipole formed by the mass anomaly at depth and the topography that it compensates.) A positive mass anomaly induces tectonic compression ($\sigma_{zz} > \sigma_{xx}$) and, vice versa, a negative mass anomaly induces extension.

Relationship (5) can be generalized to the case of a density distribution that is a function of the depth z:

$$\int_0^L \Delta\sigma_{zz}\,\mathrm{d}z = g\int_0^L \Delta\rho(z)\, z\,\mathrm{d}z. \tag{6}$$

Note that according to isostasy the topography is written as

$$h = -\int_0^L \Delta\rho(z)\,\mathrm{d}z/\rho_{\mathrm{c}}, \tag{7}$$

where ρ_{c} is the density of the crust. The two integrals in (6) and (7) do not necessarily have opposite signs (i.e. positive topography does not necessarily correspond to extension and vice versa). Such is the case in figure 2, where the shallow mass

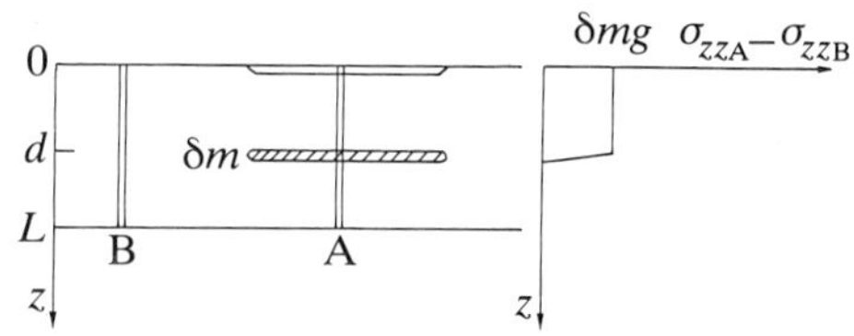

Figure 1. The mass anomaly δm, located at a depth d, induces a vertical stress increase δmg between d and the Earth's surface. $\int_0^L \sigma_{zz}\,dz$ is in turn increased by δmgd.

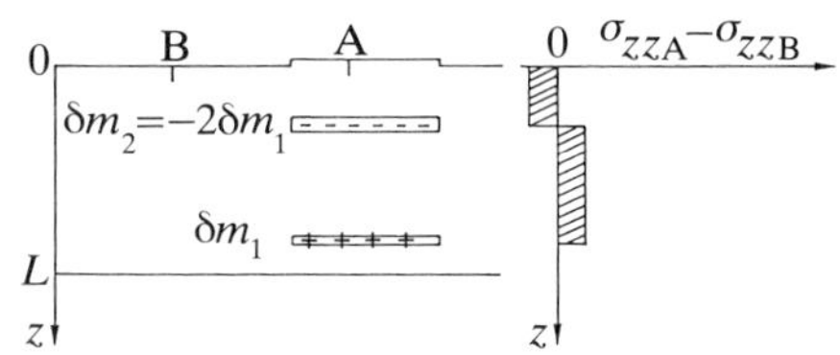

Figure 2. Despite the fact that the relief is high around A, tectonic 'compression' is stronger in A than in B. The 'moment' of the positive deep mass anomaly δm_1 is larger than the 'moment' of the negative shallower mass anomaly δm_2. Consequently, $\int_0^L(\sigma_{zzA}-\sigma_{zzB})\,dz$ (which corresponds to the difference between the shaded areas on the right of the figure) is positive.

anomaly of amplitude $-2\Delta m_1$ (Δm_1 is positive), has a dominant effect on the topography (i.e. the topography is positive), but the positive mass anomaly Δm_1, which is three times deeper, imposes a compressive tectonic style. The effect of intralithospheric mass anomalies cannot therefore be directly linked with the topography. Indeed in the next paragraph we see examples of high topography mountain chains which are in compression with respect to the surrounding, lower-altitude areas.

Note that the 'moment law' (5) is only valid for large wavelengths (i.e. when the lateral extent of the mass anomaly is large with respect to the thickness of the lithosphere). Various examples of the breakdown of this law at short wavelengths, based on the resolution of the Navier–Stokes equation, can be found in Fleitout & Froidevaux (1982).

(*b*) *Some geological examples*

High plateaus: a bound on the magnitude of deviatoric stresses in the lithosphere.

The recent tectonics of both Tibet and the Andes indicate that, when a plateau reaches an altitude of about 3000 m, the tectonic style becomes extensional. The vertical stress integrated over the whole lithospheric thickness becomes smaller (more compressive) than the horizontal stress. The excess of integrated vertical stress linked with the plateau topography and its compensation is at most of the order of 6×10^{12} Pa m (Froidevaux & Isacks 1984), which means that the integrated deviatoric stress in the moderate altitude areas surrounding the plateaus is also of the order of 6×10^{12} Pa m. Because of the geometry of continental collision and subduction in the convergence between India and Asia, the Tibet area is expected to experience one of the largest horizontal compressions in the world. The above considerations (extension in Tibet) imply that, even there, the integrated stress does not exceed 6×10^{12} Pa m. This is an important constraint on the magnitude of deviatoric stresses in the plates. Subduction in the Andes may somewhat complicate the picture.

Mountain chains in compression: a consequence of subduction of the continental lithosphere.

Plateaus correspond to the straightforward cases where high-altitude areas are in extension. On the other hand, collisional mountain chains correspond to high-altitude areas in compression, while the lower-altitude surrounding areas are often in extension. For example deformation since the Oligocene in the Massif Central has been characterized by E–W extension with sometimes N–S compression. The N–S Limagne and Forez grabens are the most prominent tectonic features there. Just east of the Massif Central, the French Alps is a compressional mountain chain undergoing E–W compression (Brereton & Müller 1991). The altitude is lower in the Massif Central than in the Alps. This is a case similar to that of figure 2; as indicated by the topography, the integrated density anomalies are more negative below the Alps than below the Massif Central. The moment of these negative densities is, however, larger for the Massif Central. Indeed the high altitude in the Massif Central is not due to crustal thickening, but to a warming up of the lithosphere and to a deep negative density anomaly with a large moment. In the Alps the altitude is due to crustal thickening, most certainly accompanied by a thickening of the lithosphere (Babuska *et al.* 1984); as in figure 2, there is a mass anomaly at shallow depth which is negative below the Alps and positive below the Massif Central. At large depths however, the mass anomalies are positive under the Alps and negative under the Massif Central. This explains the difference of tectonic style between the two areas. Note that in Europe there are numerous similar cases of compressional mountain chains accompanied by nearby extensional basins (Tuscany and Appennines, Pannonian Basin and Carpathian mountains, etc.).

Even though in nature the geometry of deformation is often more complicated, a uniform and adiabatic thickening or thinning of the lithosphere can be considered as a simplified model of these tectonic processes. It can be shown (Fleitout & Froidevaux 1982) that for standard values for a young continental platform (30 km thick crust, 100 km thick thermal lithosphere), thickened lithospheres tend to be more in compression than normal lithospheres, and thinned lithospheres are more in extension. This simply results from the fact that, if one compares a column of standard lithosphere to a column of hot mantle, the moment of the shallow negative mass anomalies due to crustal thickening is smaller (in absolute value) than the moment of the cold-temperature anomaly at depth. These two quantities may become equal for thinner lithospheres, implying also thinner or denser crusts. The moment due to the cold mass anomalies at depth is, however, rarely negligible compared with that of the crust. Simple isostatic comparison between a ridge and a continent indicate that the mantle below a continent, with an altitude close to zero and a crust 30 km thick (or thicker), is necessarily denser than the mantle below a ridge. Models of subsidence of continental basins indicate, moreover, that the continental lithosphere tends to cool and become denser as a function of time in the same manner as the oceanic lithosphere. There is therefore no reason, *a priori* to assume either the absence of lithosphere or a neutrally buoyant lithosphere below the continental crust. Large-scale models of tectonic deformation then need to deal with the not yet well-understood problem of the mechanical behaviour and stability of the thickened cold lithospheric root.

(c) How to introduce 'the moment law' in a plane-stress finite-element program?

A horizontal volume force, proportional to the horizontal gradient of the moment, just needs to be incorporated in the program. Fleitout & Froidevaux (1983) gave the equations for the case of spherical geometry. In the case of plane geometry the equations are written

$$L\left(\frac{\partial \sigma_{xx}}{\partial x}+\frac{\partial \sigma_{xy}}{\partial y}\right)=\sigma_{xz}^{L}\frac{-\partial M}{\partial x}, \quad L\left(\frac{\partial \sigma_{yy}}{\partial y}+\frac{\partial \sigma_{xy}}{\partial x}\right)=\sigma_{yz}^{L}\frac{-\partial M}{\partial y},$$

where x and y are the two horizontal directions, σ_{xz}^{L} and σ_{yz}^{L} are the resistive shear stresses due to the mantle drag below the plate and M is the moment of the lithospheric mass anomalies. Indeed, this approach has often been applied to cases where the moment is simply due to crustal thickening.

3. The relative magnitude of stresses due to 'forces acting on the plates' and to intralithospheric mass anomalies

Stresses due to intralithospheric mass anomalies are sometimes considered as a second-order effect compared with 'forces applied to the plates'. We want to emphasize here that this is not our opinion; the forces applied to the plates are of the same order of magnitude as the ridge push which is a consequence of intralithospheric mass anomalies.

(a) The ridge push: a variation of the vertical stress due to the increase of density in the lithosphere with age

The 'ridge push' is well known as a force applied to the plates and is written

$$F_{\mathrm{rp}}=\int_{0}^{L}\Delta\rho\, z\,\mathrm{d}z,$$

where $\Delta\rho$ corresponds to the density difference between an old ocean and a young ocean. It can also be written as $F_{\mathrm{rp}}=\int_0^L \Delta\sigma_{zz}\,\mathrm{d}z$. It corresponds to the difference in the integrated vertical stress due to the cooling of the lithosphere. It is evaluated to be about 2×10^{12} Pa m (Dahlen 1981; Fleitout & Froidevaux 1983).

The 'tectonic stress' in the oceanic plate depends upon the variation of $\int_0^L(\sigma_{xx}-\sigma_{zz})\,\mathrm{d}z$ as a function of age. The horizontal variation of σ_{xx} away from the ridge is given by the horizontal equilibrium

$$\frac{\partial}{\partial x}\int_{0}^{L}\sigma_{xx}\,\mathrm{d}z=-\sigma_{xz}, \tag{2}$$

where σ_{xz} is the shear stress beneath the plate. What is the sign and amplitude of σ_{xz}? In §4, it is argued that σ_{xz} is resisting the plate motion (in other words, we think passive drag to be more likely than active shear below the plates). σ_{xz} will then tend to induce an increase of tectonic extension away from the ridge. Data concerning the state of stress in the oceans indicate that the tectonic style becomes compressive at a short distance from the ridge (Stein & Pelayo 1991). This indicates that the shear stress below the plate is not large enough to overcome the compression induced by the ridge push. This puts an upper bound on this shear stress of about 2×10^{5} Pa. The evolution of the vertical and horizontal stresses away from the ridge is schematized on figure 4.

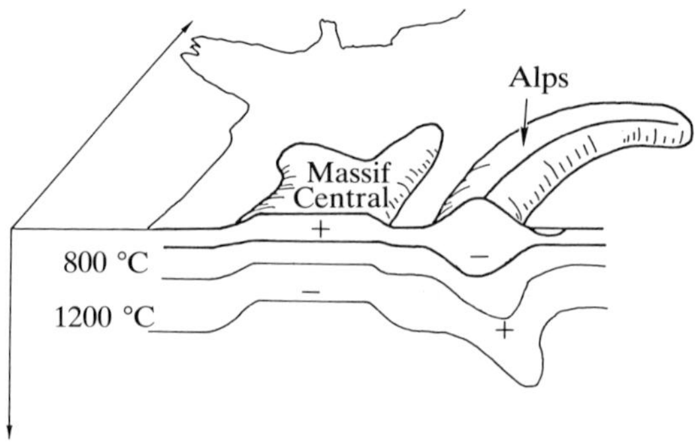

Figure 3. Schematic cross section through the Alps and the Massif Central. Crustal thickening implies lighter material at depth while the thickening of the cold (dense) mantle material forming the lower part of the lithosphere implies denser material at depth. As in figure 4, the deeper mass-anomaly has a predominant tectonic influence.

(b) *Ridge push compared with the other forces applied to the plates*

The slab pull can correspond to a very large force applied to the plates; it is of the order of 3×10^{13} Pa m. It is, however, locally counterbalanced by the colliding resistance and the resistance to penetration of the slab, so that the net force applied to the plate is not very large (Forsyth & Uyeda 1975). As argued by Bott, the fact that the state of stress is compressive in old oceanic lithospheres, close to subduction zones, even implies that the colliding resistance plus the viscous resistance to slab penetration exceed the slab pull. The sum of these forces corresponds to the deviatoric stress at old ages on figure 4.

Note that these considerations are only valid as averages and that there can be huge local variations of the force exerted by the slab depending upon the age of the subducting slab, the nature of the subducted material, the length of the slab, etc. These factors can induce large perturbations of the state of stress in the plates (Cloetingh & Wortel 1985).

(c) *The state of stress in old continental platforms*

During recent years seismic tomography has revealed thick lithospheric roots (reaching to about 400 km) beneath old continental platforms. Simple isostatic considerations imply that the seismic anomalies are not purely thermal but are also due to a petrological difference between the oceanic and continental mantles at similar depths (Jordan 1978). What about the variation of vertical stress due to such thermal and petrological features of the lithosphere? This will of course depend upon the relative distribution of the petrological and thermal anomalies as a function of depth. However, the geoid jump across passive margins is also proportional to the moment of the differential-density anomalies and thus can be used to estimate this moment. This geoid jump at passive margins with an old craton of moderate altitude amounts to about 5 m (it reaches 10 m for high continental platforms such as South Africa). This means that the moment of the density anomalies is about 10^{12} Pa m and that the averaged vertical stress is similar to that of an oceanic lithospheric about 50 Ma old (the latter is compatible with standard crustal densities, a linear temperature profile in the mantle varying between 450 °C at the base of the crust and 1300 °C at the base of the lithosphere and a petrological density-anomaly which is constant throughout the lithosphere and equal to -0.026 Mg m^{-3}). As 50 Ma old

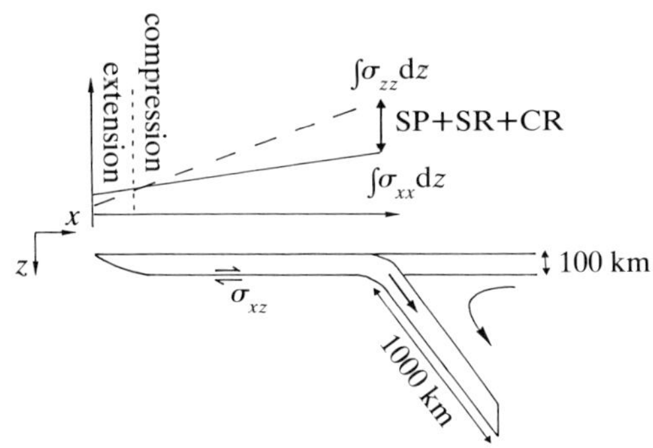

Figure 4. Simplified evolution of the horizontal and vertical stresses in an oceanic lithosphere pulled by a slab. The shear stress associated with the viscous drag beneath the plate implies an increase of the horizontal stress away from the ridge. As the lithosphere cools (i.e. becomes denser) away from the ridge, the vertical stress also increases. The 'tectonic stress' which corresponds to the difference between the horizontal and vertical stresses is observed to be extensional close to the ridge and compressive at older ages.

oceanic lithospheres are observed to be in compression, old continents are also expected to be in compression if the basal shear stress does not modify significantly the horizontal stress. Indeed most of the time they are observed to be in compression.

4. Deep density anomalies

The 'moment law' indicates that the deeper a density anomaly is in the lithosphere, the larger the tectonic stress that it induces. What then is the influence of the very deep density anomalies linked to the seismic anomalies detected during recent years by seismic tomography ? The answer to this question depends upon the structure of mantle convection.

1. For one-layer mantle convection the degree 2 and 3 density anomalies, corresponding to the seismic anomalies found in the lower mantle and responsible for the low-harmonic geoid, should induce both huge tectonic stresses in the lithosphere (larger than 10^{13} Pa m) and huge topography anomalies (of the order of ∓ 1 km) (Ricard *et al.* 1984).

2. For two-layer mantle convection the lower-mantle mass anomalies induce a deflexion of the 670 km discontinuity. However, for appropriate viscosity contrasts between upper and lower mantle, the lithosphere is found to be mechanically decoupled from the lower mantle and neither vertical stress (corresponding to topography), nor shear stress (inducing tectonic stress) are induced at the base of the lithosphere by lower-mantle mass anomalies.

No topography anomaly is observed above lower-mantle mass anomalies. This result has been obtained by an inversion of oceanic topography as a function of age plus terms proportional to the geoid (Colin & Fleitout 1990). (A new inversion has been performed without flattening terms in the expression characterizing the topography as a function of age. We got the following result: $\text{topo} = 3026 + 246\sqrt{\text{age}} - 1.1\text{geoid}^2 - 3.35\text{geoid}^3 - 1.28\text{geoid}^4 \ldots$ This still indicates very little topography associated with the low-order harmonic geoid.) Observations of variations of altitude recorded by shallow marine deposits, as the plates drift above highs and lows of the geoid, confirm this conclusion. Therefore, we prefer a two-layer

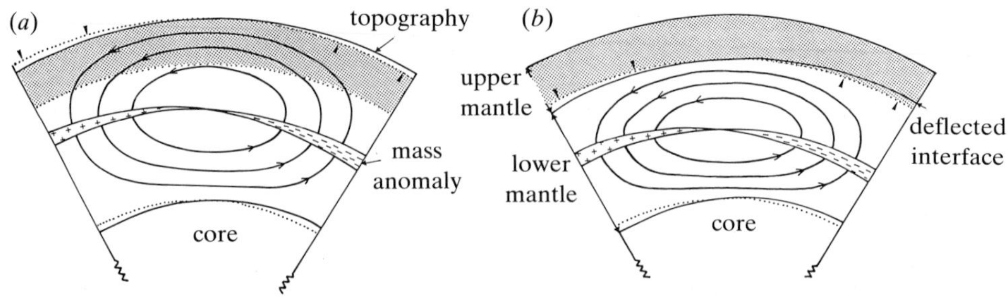

Figure 5. Influence of deep mantle density heterogeneities on the lithospheric stresses. For one-layer mantle convection (*a*), the lower-mantle mass heterogeneities induce huge lithospheric stresses and topography anomalies (of the order of 10^8 Pa × 100 km and ∓ 1 km respectively). For two-layer mantle convection (*b*), the lithosphere is mechanically decoupled from the lower mantle. The relief and the stress field are not affected by lower-mantle mass heterogeneities.

convecting mantle model where the lithosphere is mechanically decoupled from the lower mantle. Note that models where there is a convective layering of the mantle, but where the cold lithosphere can intermittently penetrate the lower mantle, would also fit these observations. In these two-layer or 'one-and-a-half-layer' mantle convection models, lower-mantle mass anomalies, which are the main source of the geoid, are decoupled from the lithosphere.

Upper-mantle seismic-tomography anomalies represent a complex mixture of chemical effects and thermal effects. It may be dangerous to interpret them simply in terms of density anomalies. Their tectonic effect should anyway be relatively minor, except for the variations in lithospheric structure discussed in §1.

5. Conclusion

What are the ingredients which must be included in a large-scale numerical model of the stresses in a plate? The lithospheric mass heterogeneities, among which are the ridge push and a resistive viscous drag below the plates, constitute the 'body forces' inside the plate. At the ridge the deviatoric stress seems to be very close to zero. The boundary condition at the trenches is more speculative. It is not yet well known how slab-pull, resistance to the slab penetration and collisional resistance vary with the geometry of the slab and the rate of subduction. Indeed, the stress field could be considered as a useful constraint on these quantities. As indicated in §4, deeper mass anomalies do not seem to play a major role in the stress field.

This paper only discussed the 'body forces' and the boundary conditions necessary for modelling and understanding the stress field. There is no doubt that the constitutive relationship relating internal deformation to the stresses is also important. In particular faults can considerably modify the stress pattern.

References

Artyushkov, E. V. 1973 Stresses in the lithosphere caused by crustal thickness inhomogeneities. *J. geophys. Res.* **78**, 7675–7708.

Babuska, V., Plomerava, J. & Sileny, J. 1984 Large scale orientated structures in the subcrustal lithosphere of Central Europe. *Annls Geophysicae* **16**, 117–123.

Brereton, R. & Müller, B. 1991 European stress: contributions from borehole breakouts. *Phil. Trans. R. Soc. Lond.* A **337**, 165–179. (This volume.)

Cloetingh, S. A. P. L. & Wortel, M. J. R. 1985 Regional stress field of the Indian plate. *Geophys. Res. Lett.* **12**, 77–80.

Colin, P. & Fleitout, L. 1990 Topography of the ocean floor: thermal evolution of the lithosphere and interaction of deep mantle heterogeneities with the lithosphere. *Geophys. Res. Lett.* **17**, 1961–1964.

Dahlen, F. A. 1981 Isostasy and the ambient state of stress in the oceanic lithosphere. *J. geophys. Res.* **86**, 7801, 7807.

Fleitout, L. & Froidevaux, C. 1982 Tectonics and topography for a lithosphere containing density heterogeneities. *Tectonics* **1**, 21–56.

Fleitout, L. & Froidevaux, C. 1983 Tectonic stresses in the lithosphere. *Tectonics* **2**, 315–324.

Forsyth, D. & Uyeda, S. 1975 On the relative importance of the driving forces of plate motion. *Geophys. Jl R. astr. Soc.* **43**, 163–200.

Froidevaux, C. & Isacks, B. L. 1984 The mechanical state of the lithosphere in the Altiplano-Puna segment of the Andes. *Earth planet. Space Lett.* **71**, 305–314.

Jordan, T. H. 1978 Composition and development of the continental tectosphere. *Nature, Lond.* **274**, 544–548.

Ricard, Y., Fleitout, L. & Froidevaux, C. 1984 Geoid heights and lithospheric stresses for a dynamic Earth. *Annls Geophysicae* **12**, 267–286.

Stein, S. & Pelayo, A. 1991 Seismological constraints on stress in the oceanic lithosphere. *Phil. Trans. R. Soc. Lond.* A **337**, 53–72. (This volume.)

Discussion

P. England (*Oxford University, U.K.*) I would disagree that the continents are negatively buoyant because of deep cold roots because Dr Fleitout's results are very sensitively dependent on the assumed distribution of density with depth and, secondly, upon the assumption that lithosphere is preserved during the deformation. I would like to argue from her data that in fact the lithosphere is not preserved during compressional deformation. I noticed that the big negative velocity anomalies were concentrated under the Aegian, Turkey and the Caucasus which respectively are undergoing compression, strike–slip faulting and compression. So there does not seem to be a very strong correlation between upper-mantle thermal state and tectonic style. Secondly, Romanowicz and co-workers demonstrated that the thickness of lithosphere under Tibet is very much less than you would expect than if you had simply, homogeneously shortened the lithosphere layer by a factor of two. Although I would agree with her that undeformed continental lithosphere appears to be negatively buoyant with respect to the mid-ocean ridges it appears that once you start compressing that lithosphere it doesn't stay negatively buoyant for some reason.

N. Kusznir (*Liverpool University, U.K.*). If you want to get rid of the root by thermal assimilation you could explain why orogeny is going through extension but if you get delamination it falls off and you still have a negatively buoyant zone down there which should be subject to the same physics in terms of the density contrast moment. Although it may not be physically attached to the lithosphere above it should still have an effect. Indeed if one takes the density contrast moment into account the deeper it goes the greater the compression. One somehow senses one is missing something from the physics here.

Sublithospheric loading and plate-boundary forces

By M. H. P. Bott
Department of Geological Sciences, University of Durham, South Road, Durham DH1 3LE, U.K.

Sublithospheric loading arises from anomalous densities in the mantle, such as cool subducting slabs and regions affected by hot spots. Such loading gives rise to isostatic flexure and to tectonic stress in the strong upper lithosphere. Simple models of sublithospheric loading have been studied by finite-element analysis. The maximum loading stress produced by a simple load increases with its width towards the theoretical density-moment function value, but is found to be almost independent of depth for a narrow load (in contrast to a wide load). A layer of low viscosity above the load reduces the stress, depending on its thickness and viscosity.

If the lithospheric stress arising from loading is intersected by a zone or plane of weakness, then plate-boundary tractions develop on the adjacent plates resulting from the redistribution of stress. It is shown, by modelling, that ridge push, slab pull and trench suction can be explained in this way.

---

1. Introduction

Tectonic stress in the lithosphere must be produced by a source which in general can be renewed as rapidly as it is dissipated by ongoing tectonic activity. The most obvious such source is sublithospheric loading associated with regions of anomalous density in the mantle. The best-known sources of this type include the hot low-density upwelling asthenosphere beneath ocean ridges, the upper mantle hot spots beneath continental and oceanic swells, and cool dense subducting oceanic lithosphere. This paper examines how sublithospheric loading produces tectonic stress in the strong upper lithosphere, and how such stress systems can give rise to plate-boundary forces.

Subsurface loading is usually the primary load in active tectonic regions and may occur within and/or below the lithosphere. The isostatic flexural response of the lithosphere to the deep load produces a secondary topographic load at the surface. Loading stress is produced by a combination of such subsurface and surface loading. Stresses produced by loading are of two types. *Bending stress* results from local asymmetry between the surface and subsurface loads. *Loading stress* without bending occurs when surface and subsurface loads mirror-image each other as in simple local isostatic equilibrium. Flexural isostatic equilibrium gives rise to superimposition of both loading and bending stress. Loading stress concentrates upwards into the strong upper lithosphere and, in contrast to bending stress, its magnitude is inversely proportional to the thickness of the strong upper lithosphere (Kusznir & Bott 1977).

The concept of local isostatic loading stress was discussed qualitatively by Bott (1971, p. 237) and subsequently evaluated across passive margins (Bott & Dean 1972), for more general crustal thickness variations (Artyushkov 1973) and for sublithospheric loading (Bott & Kusznir 1979). For structures which are wide

Phil. Trans. R. Soc. Lond. A (1991) **337**, 83–93
Printed in Great Britain

compared with their depth an approximate estimate of the deviatoric stress σ'_{xx} in an elastic layer of thickness T produced by local isostatic loading can be obtained using the density-moment function (Parsons & Richter 1980; Dahlen 1981)

$$\sigma'_{xx} = \frac{1}{2T}\int_0^D gz\Delta\rho\,\mathrm{d}z, \tag{1}$$

where z is depth below surface, D is depth to base of anomalous density $\Delta\rho$ and g is gravity. The geoid anomaly Δh of a wide structure in local isostatic equilibrium is also proportional to the density-moment function (Haxby & Turcotte 1978)

$$\Delta h = \frac{-2\pi G}{g}\int_0^D z\Delta\rho\,\mathrm{d}z. \tag{2}$$

Thus for very wide loads, the deviatoric loading stress is proportional to the geoid anomaly, and both increase linearly with the depth of the load.

This paper models loading stress in the strong upper lithosphere and associated flexural topography produced by sublithospheric loading, using elastic/viscoelastic finite-element modelling based on the initial strain method. The finite-element package has been developed in the Durham department, with the most substantial input being made by Waghorn (1984). Triangular and quadrilateral isoparametric elements with six and eight nodes respectively are used. Faults can be included using the dual node technique (Goodman *et al.* 1968). The models are run for a sufficient number of time steps for relaxation of the stress in the most viscous region, with time steps being short enough to avoid instability in the least viscous region. The process of approaching dynamic equilibrium in the models is not of geological significance (as it would be for post-glacial uplift studies) as the geological timescale of development of major tectonic structures is much longer.

The modelling assumes the out-of-plane deviatoric stress is zero, as this is most suitable for the viscoelastic creep computation. However, the elastic layer deformation and the in-plane stress differences are almost identical to results obtained with the plane strain formulation. Both the stresses and the flexural uplift are computed. The geoid anomaly can also be computed from the calculated free air gravity anomaly of the models using the fast Fourier transform (FFT) to carry out the transformation in the wavenumber domain. The anomalous densities are referenced to a specified density–depth distribution such as at an edge of the model. The computed stresses can thus be regarded as being superimposed on this standard lithostatic pressure distribution.

2. Simple model of sublithospheric loading

A simple symmetrical two-dimensional finite-element grid extending 3000 km horizontally and down to 700 km depth, neglecting the Earth's curvature, has been used to study sublithospheric loading. An upper elastic layer 25 km thick overlies a substratum with viscosity of 10^{21} Pa s. The resulting stresses, however, are independent of the viscosity although the relaxation time is proportional to viscosity. Young's modulus and Poisson's ratio are 1.75×10^{11} Pa and 0.27 respectively throughout. The flexural rigidity of the elastic layer is 2.46×10^{23} N m which is near the middle of the continental range (Walcott 1970). For the purpose of applying isostatic restoring forces, a mantle density of 3330 kg m^{-3} has been assumed. The load consists of a 100 km × 100 km region of square cross section with

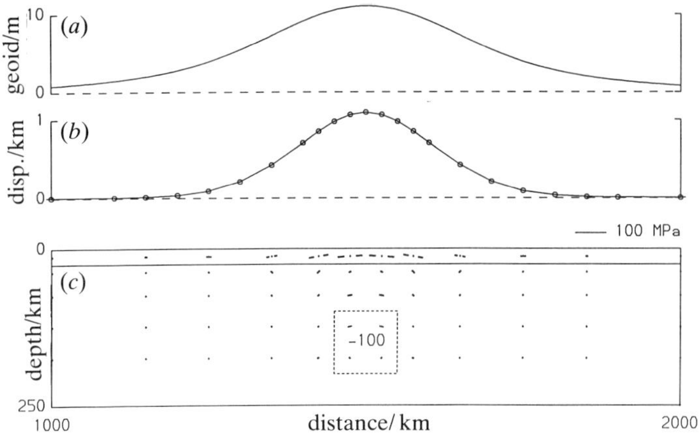

Figure 1. The deviatoric stress distribution (*c*), surface elevation profile (*b*) and geoid anomaly (*a*) at 20 km height resulting from the two-dimensional $-100\ \text{kg m}^{-3}$ region of anomalously low density outlined by the broken line. The solid line denotes the base of the 25 km thick elastic layer. The in-plane deviatoric stress nearest to the horizontal is shown, with a thick broken line denoting tension and (in other figures) a thick unbroken line denoting compression.

an anomalous density of $-100\ \text{kg m}^{-3}$ relative to the normal mantle, situated beneath the centre of the symmetrical model. The stresses and surface displacements would be equal but of opposite sign for an equivalent positive density anomaly. The average deviatoric stress nearest to the horizontal is shown at the centre of each element (the other in-plane deviatoric stress is equal but opposite). Thus the small bending stresses are eliminated as the elastic layer is one element thick.

Model SLLA (figure 1), with the centre of the load at 150 km depth, has been run for 200 time steps of 500 years each (totally 100000 years) to enable the shape of the uplift produced by the load to develop almost fully. The main features are (1) a flexural uplift reaching a maximum of 1028 m, (2) a free air geoid anomaly of 11.05 m maximum relative to the value at the distant edges, and (3) deviatoric stresses which have relaxed to small values in the viscous region but have concentrated up into the strong elastic layer reaching a maximum value of 78 MPa at the centre. This model demonstrates in a simple way the effectiveness of a sublithospheric load in causing vertical isostatic displacement at the surface and in stressing the strong upper lithosphere. The timescale for this to occur is much shorter than that required to dissipate the load by advection. However, because of the finite lateral extent of the load, the geoid maximum is only 17% of the theoretical value from equation (2) and the maximum deviatoric stress is only 27% of the theoretical value from equation (1).

Figure 2 illustrates how the deep load in figure 1 couples to the strong lithosphere. The load gives rise to an anomalous pressure field above it which acts on the base of the elastic layer to produce the flexural uplift. Part of the deviatoric stress in the layer is produced by this pressure and that of the surface topography, but most of it arises from the outwardly directed shear traction exerted by the advecting viscous material driven by the load.

The variation of maximum height, maximum geoid anomaly and maximum deviatoric stress is shown in figure 3*a* as a function of width/depth ratio, with the

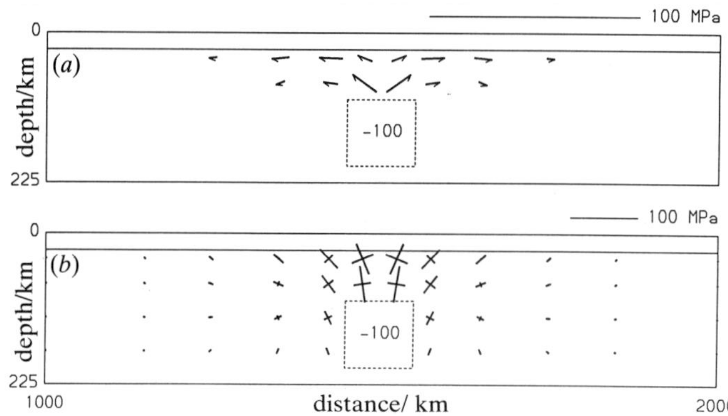

Figure 2. The element-averaged compressional principal stress (*b*) and shear stress (*a*) in the viscoelastic region beneath the elastic layer for Model SLLA (figure 1), demonstrating the anomalous normal pressure and shear stress applied to the base of the elastic layer as a result of the deep load.

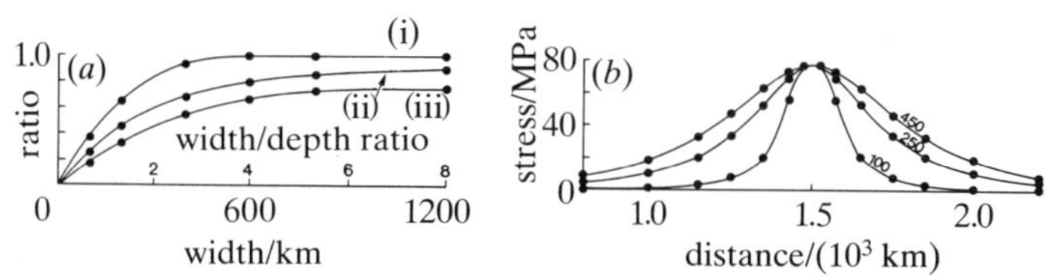

Figure 3. (*a*) The maximum values of deviatoric stress, surface elevation and geoid anomaly as ratios to the values for an infinitely wide structure, plotted against width (and width/depth ratio) for a 100 km thick load centred at 150 km depth as in Model SLLA, but of varying width. (i) Height; (ii) stress; (iii) geoid. (*b*) The profile of element-averaged deviatoric stress in the elastic layer for models similar to SLLA but with varying depth to the centre of the load. Depth is shown in kilometres above the profiles.

load of Model SLLA being progressively widened. These loads are all centred at 150 km depth and are 100 km thick. Figure 3*a* shows that the maximum elevation increases rapidly with width/depth ratio towards the local isostatic value, and the maximum deviatoric stress and geoid anomaly tend rather more gently towards the theoretical values given by equations (1) and (2). The effect of varying the depth of the simple square load of Model SLLA on the distribution of stress is shown in figure 3*b*. The variation of load depth has almost insignificant effect on the maximum stress value. The width of the stressed region increases approximately linearly with load depth, as might be anticipated.

The effect of a low viscosity layer above the load has been explored in a preliminary way. (i) Model SLLA was modified by reducing the viscosity of the load by a factor of 100. This has a negligible effect. (ii) a 25 km thick layer of reduced viscosity was additionally incorporated immediately above top of the load. For a viscosity reduced by a factor of $\times 100$ the maximum deviatoric tension is 56 MPa and for reduction by $\times 400$ it is 49 MPa (a reduction of about 35%). For loading at this sort of depth, typical for an upper mantle hot spot, the reduction in the stress appears to be significant but not overwhelming. (iii) a reduction of viscosity by a factor of $\times 100$ was next applied to a layer above a deep-seated load centred at 350 km depth. The layer was varied in thickness from 25 to 225 km, with the stress in the upper elastic lithosphere being reduced by about 40–45% in all the models.

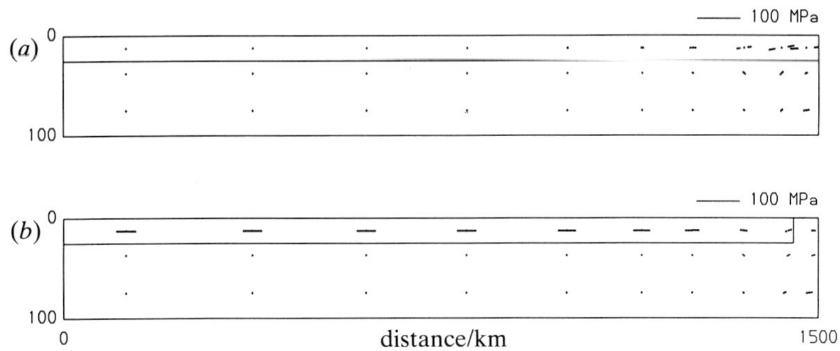

Figure 4. Comparison of the deviatoric stresses in Model SLLA (*a*) with those developed when the model is modified in (*b*) by including a weak crest as shown and constraining the distant edge nodes of the elastic layer to zero horizontal displacement. Note that only one half of the symmetrical models, from the far left edge to the centre, is shown. Vertical exaggeration is ×2.

3. Loading stress and plate-boundary force

Large loading stress might be expected to occur at plate boundaries where prominent sublithospheric loading occurs, such as low-density asthenospheric upwelling beneath ocean ridges and dense sinking slabs beneath subduction plate boundaries. If these boundaries could be strong enough to withstand the loading stress, ridges would be associated with large local extensional stress distributions and subduction zones with large compressions. How, then, do the plate-boundary forces of apparently opposite polarity develop at these boundaries?

This is explained in principle by reference to Model SLLA and its modification shown in figure 4. This diagram shows the shallower part of the left-hand half of the symmetrical models out to their left edge. In figure 4*a* Model SLLA has a continuous elastic layer as in figure 1. In figure 4*b* the model is modified in two ways. First, the central two elastic elements of the symmetrical model have been replaced by viscous elements extending up to the surface to simulate a weak crest. Second, the left edge nodes of the elastic layer have been constrained to zero horizontal displacement.

If the edge is fixed without weakening the crest (not shown), then a small compression occurs out to the fixed edge as expected. If the crestal region is also weakened as shown, then the tensions in the vicinity of the crest are greatly reduced and large compression occurs beyond the central region and extends out to the fixed edge. The effect of weakening the crest is to redistribute the stress throughout the elastic plate by superimposing a large supplementary compression which is slightly smaller than the maximum tension in the unweakened model. The small residual tensions in the vicinity of the weakened crest are associated with the ongoing viscous deformation at the crest, and can be regarded as the plate-boundary resistance. The low-density load thus gives rise to a compressional stress slightly smaller than the local extensional stress of the unweakened model. This extends out to the fixed edge of the plate. If there is viscous drag on the underside of the elastic plate, then this compressional stress will decrease with increasing distance from the plate boundary.

The stress redistribution results from the modified normal pressure acting on the vertical edge of the elastic plate adjacent to the weak element. The plate boundary force, however, can be regarded as arising partly from this increased pressure and partly from the shear traction acting on the base of the nearby part of the elastic

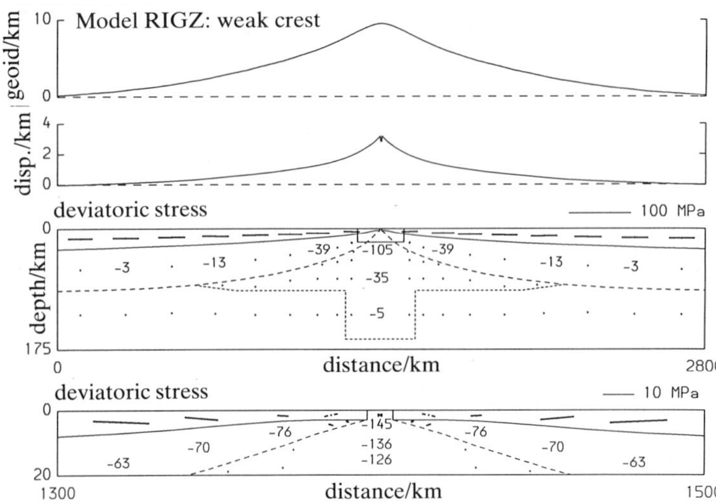

Figure 5. A model of a symmetrical ocean ridge spreading at 15 mm a^{-1} with loading of thermal origin below and within the cooling lithosphere (see text), showing deviatoric stress (enlarged central region below), flexurally developed surface elevation, and geoid anomaly at 20 km height. Anomalous density values as shown are in kg m^{-3} and are referenced to the values at the same depth beneath the old ocean floor at the edges. The base of the elastic lithosphere is marked by the solid line and the base of the seismological lithosphere by the coarse broken line. Vertical exaggeration is $\times 3$ for the full length section.

layer as shown in figure 2. For a deep load such as in Model SLLA, the shear traction effect is strongly dominant.

The plate boundary force, referenced to the normal density–depth profile at the edge, can be determined by integrating the stress difference across the thickness of the elastic layer at the edge. The deviatoric compressional stress at the left edge is 46 MPa, corresponding to a plate-boundary force of 2.3×10^{12} N m^{-1}.

Ridge push

Model RIGZ (figure 5) illustrates the origin of the ridge push force. This model is of a type discussed by Bott (1991), to which reference should be made for further detail. It represents a symmetrical ocean 2800 km wide spreading at 15 mm a^{-1}. The grid extends to 400 km depth with the basal nodes being free. The lithospheric structure is based on the cooling plate model of Parsons & Sclater (1977) and the elastic thickness model of Watts (1978). The thicknesses of the elastic upper lithosphere and the high-viscosity lower lithosphere (10^{23} Pa s) increase from the ridge crest accordingly. The asthenosphere below has an assumed viscosity of 10^{21} Pa s. The density structure as inferred from the temperature field is referenced to 90 Ma oceanic lithosphere at the edges. The uniform oceanic crust is not included, so that the density is taken to increase by 2300 kg m^{-3} at the sea floor for isostatic purposes.

The weak crest is simulated by assigning an asthenospheric viscosity to the two adjacent shallowest elements. The enlargement of the crestal region (below in figure 5) shows that there is a small subhorizontal deviatoric tension near the crest, representing ridge resistance. The stresses become compressional at a small distance from the crest, rapidly increasing to nearly 40 MPa which is maintained to the edges.

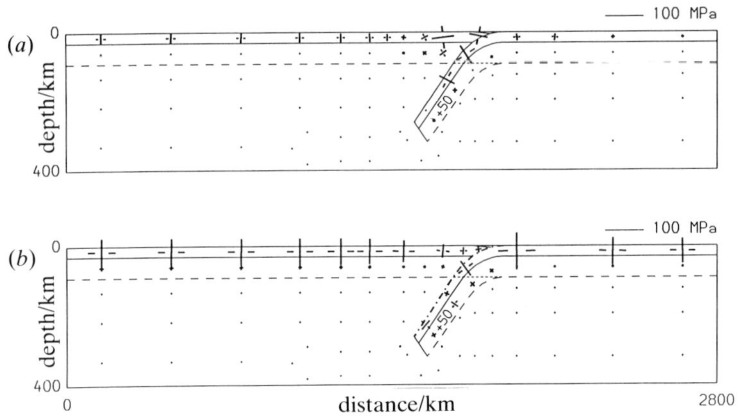

Figure 6. Deviatoric stress distributions produced by (*a*) Model SLA1 (subduction fault locked) and (*b*) Model SLA2 (subduction fault free), with a 45° slab extending to about 300 km depth. Details as in figure 5 and as described in the text. Vertical exaggeration is $\times 1.5$.

The compression is approximately constant over most of the flanking regions because of the trade-off of increasing elastic layer thickness with increasing tectonic force.

The model develops an isostatic elevation of 3.2 km relative to the old ocean floor. The geoid maximum relative to the old ocean is 9.5 m, compared to the theoretical value of 11.7 m calculated for an infinitely wide structure using equation (2). The ridge-push force developed at the fixed edges is 2.43×10^{12} N m^{-1} compared with the theoretical value of 2.69×10^{12} N m^{-1} using equation (1). The discrepancies between the modelled and theoretical values of the stress field and the geoid are much smaller than those discussed in the previous section of the paper because the anomalous density structure is much wider compared with its depth. The ridge resistance slightly reduces the plate-boundary force.

The above estimate of ridge-push force is in good agreement with other estimates (Dahlen 1981; Fleitout & Froidevaux 1983). Bott (1991) has shown that about double this ridge-push force should develop as a result of such a ridge being underlain by an anomalously hot upper mantle with an average temperature anomaly of 70 K down to 400 km depth, which is typical for a hot spot region such as parts of the North Atlantic.

Slab pull and trench suction

Loading stresses associated with subduction and the related plate-boundary forces have been modelled by Bott *et al.* (1989) and by Whittaker *et al.* (1991). Two of the models shown by Whittaker *et al.* (1991) have been replotted in horizontally extended form in figure 6. The model is 2800 km wide and extends to 650 km depth, with vertically fixed nodes at the base. The 45° dipping slab extends to nearly 300 km depth and has an anomalous density of +50 kg m^{-3} representing a temperature about 500 K lower than that of the adjacent normal upper mantle. The nodes at both edges of the lithosphere are constrained to zero horizontal displacement so that the force exerted by the subduction system can be determined.

The 90 km thick lithosphere consists of an upper 30 km thick elastic layer overlying a stiff lower part with viscosity of 5×10^{22} Pa s. Oceanic crust is not included, but there is 30 km thick continental crust on the overriding plate with a deficient density of -400 kg m^{-3} relative to the mantle, in local isostatic equilibrium with elevated topography relative to the ocean floor which is simulated by a surface load. In the continental crust, Young's modulus is 0.9×10^{11} Pa and Poisson's ratio is 0.25. Otherwise the mechanical properties are as in the previous models except that viscosity is higher by a factor of ten below 400 km representing a downward increase at the mantle transition zone. The upper surface of the subducting slab is a fault which can be either locked or unlocked. Further details of the models are given by Whittaker *et al.* (1991).

Model SLAB1 shows the deviatoric stresses which develop after relaxation over 250000 years when the subduction fault is locked. The slab is approaching equilibrium apart from ongoing rollback. Downdip deviatoric tension occurs in the slab, dying off to about zero at the base. The overriding plate above the slab is characterized by large subhorizontal deviatoric compressions of up to about 70 MPa caused by the sublithospheric loading of the dense slab. A small deviatoric tension of 25 MPa extends to the edge, representing the loading stress of the continental crust. The compressional stresses associated with sublithospheric loading die off away from the region underlain by the slab and are small at both edges of the lithosphere.

Model SLAB2 shows the effect of unlocking the fault, which is assigned zero friction. The slab is still sinking and rolling back in this model, and the resistance at the base gives rise to downdip compression which gives way upwards to downdip tension, as is commonly observed in some subducting slabs (Isacks & Molnar 1971). The effect of separating the two plates by a plane of weakness represented by the subduction fault is to superimpose an extensional stress which extends out to the fixed edges of both plates. The deviatoric tension at the edge of the subducting plate is 65 MPa, and at the edge of the overriding plate it is 70 MPa including the 25 MPa from continental crustal loading.

In Model SLAB2, the subducting system is pulling both plates towards the subduction zone, causing plate interior tension in this isolated situation. Integrating the stresses across the lithosphere at the edges of the model, after allowing for the continental crustal loading stress, yields estimates of the plate-boundary forces referenced to old ocean floor as follows. The slab pull force is -4.8×10^{12} N m^{-1} and the trench suction is -4.2×10^{12} N m^{-1}, these being of comparable magnitude as might be expected from the equilibrium of the system.

Whittaker *et al.* (1991) additionally studied vertically dipping slabs, slabs extending to 400 km depth, and the influence of a back-arc basin. In general, the dip of the slab does not greatly affect the plate-boundary forces but they increase with increasing depth extent of the slab. These models all show that slab pull and trench suction originate from the intersection of the compressional loading stress caused by the dense sinking slab by the weak subduction fault. The modification to the boundary tractions on both sides of the fault plane required to annihilate the shear stress gives rise to both local vertical deformation associated with the fault and to the pervasive tensional stresses which extend into the plate interiors.

4. Summary and conclusions

A simple model of sublithospheric loading consisting of a 25 km thick elastic layer above a uniform viscous substratum has been studied by elastic/viscoelastic finite-element analysis. The load is a 100 km × 100 km region of anomalously low density initially centred at 150 km depth. This load causes flexural isostatic uplift and horizontal extensional deviatoric stress in the elastic layer. The stress resulting from the loading is applied to the base of the elastic layer by a combination of excess normal pressure above the load and outwardly directed viscous drag exerted by the advecting viscous material.

Widening the load causes the deviatoric loading stress and the geoid anomaly to increase towards the values given by the density-moment function. Deepening the compact square load has an insignificant effect on the maximum deviatoric stress but the stressed region widens approximately proportionally to the depth of the load. A two-order of magnitude reduction of viscosity of a layer above the load reduces the deviatoric stress by up to 50%.

If the two central elastic elements are made viscoelastic and the edge elastic nodes are fixed horizontally, the sublithospheric loading gives rise to much reduced tension in the crestal region and compression beyond the region of the load. A plate-boundary force akin to ridge push develops resulting from the redistribution of stress. It is demonstrated how this principle applies to a thermal model of a slow spreading ocean ridge, with weak tension occurring beneath the crest and deviatoric compression of about 40 MPa beneath the older ocean floor. The ridge push referenced to old ocean floor is 2.4×10^{12} N m^{-1} and the geoid anomaly is 9.5 m, these being only 10% and 20% respectively below the density-moment function values because the load is wide compared to its depth.

A model of a 45° dipping dense subducting slab extending to about 300 km depth produces large near-horizontal compressional deviatoric stresses in the overriding plate above the slab when the subduction fault is locked. When the fault is unlocked, radical redistribution of stress occurs in the elastic upper lithosphere, with tension developing out to the plate edges when these are fixed horizontally. The slab pull force is determined as -4.8×10^{12} N m^{-1} and the trench suction force as -4.2×10^{12} N m^{-1} for this model.

The computations and manuscript have been produced using the facilities of the Durham University Computer Centre, for which the author is grateful.

References

Artyushkov, E. V. 1973 Stresses in the lithosphere caused by crustal thickness inhomogeneities. *J. geophys. Res.* **78**, 7675–7708.

Bott, M. H. P. 1971 *The interior of the Earth.* London: Edward Arnold.

Bott, M. H. P. 1991 Ridge push and associated plate interior stress in normal and hot spot regions. *Tectonophys.* (In the press.)

Bott, M. H. P. & Dean, D. S. 1972 Stress systems at young continental margins. *Nature, Lond.* **235**, 23–25.

Bott, M. H. P. & Kusznir, N. J. 1979 Stress distributions associated with compensated plateau uplift structures with application to the continental splitting mechanism. *Geophys. Jl R. astr. Soc.* **56**, 451–459.

Bott, M. H. P., Waghorn, G. D. & Whittaker, A. 1989 Plate boundary forces at subduction zones and trench-arc compression. *Tectonophys.* **170**, 1–15.

Dahlen, F. A. 1981 Isostasy and the ambient state of stress in the oceanic lithosphere. *J. geophys. Res.* **86**, 7801–7807.

Fleitout, L. & Froidevaux, C. 1983 Tectonic stresses in the lithosphere. *Tectonics* **2**, 315–324.

Goodman, R. E., Taylor, R. L. & Brekke, T. L. 1968 A model for the mechanics of jointed rock. *J. Soil Mech. Fdns Div. Am. Soc. civ. Engrs* **94**, 637–658.

Haxby, W. F. & Turcotte, D. L. 1978 On isostatic geoid anomalies. *J. geophys. Res.* **83**, 5473–5478.

Isacks, B. & Molnar, P. 1971 Distribution of stresses in the descending lithosphere from a global survey of focal-mechanism solutions of mantle earthquakes. *Rev. Geophys.* **9**, 103–174.

Kusznir, N. J. & Bott, M. H. P. 1977 Stress concentration in the upper lithosphere caused by underlying visco-elastic creep. *Tectonophys.* **43**, 247–256.

Parsons, B. & Richter, F. M. 1980 A relation between the driving force and geoid anomaly associated with mid-ocean ridges. *Earth planet. Sci. Lett.* **51**, 445–450.

Parsons, B. & Sclater, J. G. 1977 An analysis of the variation of ocean floor bathymetry and heat flow with age. *J. geophys. Res.* **82**, 803–827.

Waghorn, G. D. 1984 *Numerical modelling of the stress regime at subduction zones.* Ph.D. thesis, University of Durham, Durham.

Walcott, R. I. 1970 Flexural rigidity, thickness, and viscosity of the lithosphere. *J. geophys. Res.* **75**, 3941–3954.

Watts, A. B. 1978 An analysis of isostasy in the world's oceans. 1. Hawaiian–Emperor seamount chain. *J. geophys. Res.* **83**, 5989–6004.

Whittaker, A., Bott, M. H. P. & Waghorn, G. D. 1991 Stresses and plate boundary forces associated with subduction plate margins. *J. geophys. Res.* (In the press.)

Discussion

S. Stein (*Northwestern University, Evanston, U.S.A.*). Has Professor Bott tried to couple the continental-rift-style model with the zone of extension because the extension zone in East Africa is about a 1000 miles across as opposed to about 10 km for a mid-ocean ridge?

M. H. P. Bott. In East Africa the eastern and western rift systems are narrow and are separated by a wide cratonic region which is not undergoing extension.

R. Wortel (*Utrecht University, The Netherlands*). If I remember correctly the model of the Alps with the root was proposed by Müller and colleagues on the basis of seismic velocity information which they inferred from the propagation of seismic surface waves and I am not aware of any information about the root being detached from the surface.

M. H. P. Bott. The deep lithospheric root beneath the Alps has been inferred from both seismological and gravity evidence, and is also implied by the plate convergence. It appears to be attached to the lithosphere above but in reality is presumably separated from the strong upper lithosphere by ductile lower lithosphere as in the models.

N. Kusznir (*Liverpool University, U.K.*). Has Professor Bott modelled the net effect of the transition from continent to ocean where there is a substantial density contrast at depth? Is there any consensus from geoid information as to what happens as one goes from continent to ocean?

M. H. P. Bott. The ocean ridge models shown do not include a denser subcontinental mantle, although such may exist and would have the effect of increasing the compressional stress in the continental region. This should be reflected in the geoid anomaly, and L. Fleitout, who has worked on this problem, may wish to comment further.

L. Fleitout (*École Normale Supérieure, Paris, France*). I think it depends on the particular passive margin. I think the first work in this field was done by Turcotte and others who commented that there was no change in the geoid as one crossed passive margins. I have looked at that problem recently and it is true that on many, though not all passive margins, there is not a visible geoid jump across the boundary.

S. Stein (*Northwestern University, Evanston, U.S.A.*). On a lot of passive margins people have estimated that the flexural stresses that one gets from the sediments are enormous and I am curious to know how rapidly these sorts of forces would dissipate. Has Professor Bott calculated how these forces would drop off towards the interior of a continent?

N. Kusznir (*Liverpool University, U.K.*). The bending stresses due to sedimentation at continental margins are very large and calculations suggest that brittle failure should occur. I think there is evidence for the large bending stresses and brittle failure because the effective elastic thicknesses that we see for such systems are very small and are quite unlike, for example, ocean basin thicknesses. I think one of the reasons the effective elastic thicknesses are so thin is that bending stresses have generated failure. Why don't we get earthquakes? Could this be due to the fact that bending stresses are non-renewable?

M. H. P. Bott. I agree that the non-renewable nature of bending stresses may account for the lack of earthquakes associated with plate interior bending stress.

M. L. Zoback (*US Geological Survey, Menlo Park, U.S.A.*). There is some relevant data. We know the orientation of the margins so presumably we can know the orientation of the extensional stresses due to sediment loading and the extensional stresses due to the density contrast between crust and mantle.

J. Cartwright (*Imperial College London, U.K.*). To get the stresses necessary for extension on a continental-rift scale in your models, with the wavelength you showed, you always had a topographic effect of between 1 and 2 km. Is there any situation that can be envisaged for major continental rifts without that topographic effect on their shoulders? If there is a flat borderland to the rift, what would be the stresses for that sort of condition?

M. H. P. Bott. Yes, the rift systems of interplume passive margins and of the North Sea differ from the uplifted Tertiary continental rift systems in being formed at low elevation. Extensional stress of distant origin, such as from plate-boundary forces, was presumably the cause of these low-lying rift systems, in contrast to the main Tertiary rift systems.

The distribution of stress with depth in the lithosphere: thermo-rheological and geodynamic constraints

By N. J. Kusznir

Department of Earth Sciences, University of Liverpool, Liverpool L69 3BX, U.K.

Horizontal tectonic forces arising from the plate-tectonic process give rise to horizontal stress within the lithosphere whose depth distribution is controlled by plastic and brittle strength. A constant force model of lithosphere deformation incorporating the brittle–viscoelastic response of each infinitesimal component of lithosphere allows for the transient computation of lithosphere stress and overcomes the steady-state approximation of the constant strain-rate model. For the constant force model, creep in the lower crust and mantle leads to stress decay in these regions and to stress amplification in the upper lithosphere through stress redistribution.

The lithosphere stress–depth relationship is primarily controlled by rheology which depends on geothermal gradient, crustal composition and thickness. For continental lithosphere, low-strength regions exist within, and at the base of, the crust due to vertical contrasts in composition and therefore rheology. Changes in tectonic force with time give reversals in the sign of stress with depth so that lithosphere has a 'memory' of tectonic force history. Continental lithosphere strength calculated by the constant force model is comparable with estimated levels of intraplate tectonic force.

Horizontal stress within oceanic lithosphere arises due to plate-boundary forces and cooling of oceanic lithosphere. Application of the increasing ridge-push force with age to a cooling thermo-rheological model of oceanic lithosphere predicts compressive horizontal stresses whose magnitude increases with age and which decrease almost linearly with depth. A thermo-rheological model of thermal stress for cooling oceanic lithosphere, incorporating plastic and brittle deformation, predicts horizontal compression in the upper lithosphere above deeper tension. Cooling and ridge push combine to produce a compressive stress field in the ocean basins capable of producing brittle failure.

1. Introduction

The large lateral density inhomogeneities associated with constructive and destructive plate-boundary processes, with isostatically compensated plateau uplifts and intraplate hotspots generate horizontal tectonic force within the lithosphere (Fleitout & Froidevaux 1982, 1983; Bott & Kusznir 1984; Fleitout 1991; Bott 1991). Within intraplate lithosphere this horizontal tectonic force, F_x is distributed vertically as stress to give the local non-lithostatic stress depth distribution such that

$$F_x = \int_0^L \sigma_x \, \mathrm{d}z,$$

Phil. Trans. R. Soc. Lond. A (1991) **337**, 95–110
Printed in Great Britain

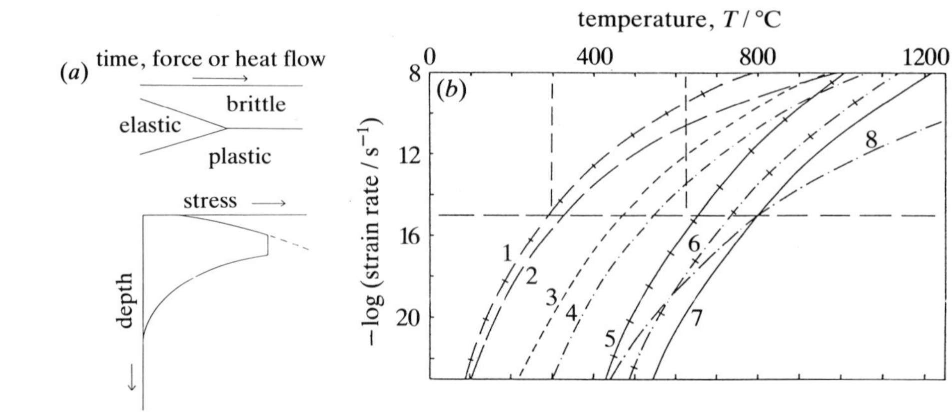

Figure 1. (*a*) Diagrammatic representation of regions of brittle, elastic and ductile deformation within the lithosphere and their control on the stress–depth distribution. (*b*) Strain rate plotted logarithmically against temperature for a range of rocks and minerals relevant to lithosphere deformation. Curves assume a stress difference ($\sigma_1 - \sigma_2$) of 5 MPa and dislocation creep. Data from Ave Lallement (1978), Goetze (1978), Koch *et al.* (1980), Kohlstedt & Goetze (1974), Post (1977), Shelton & Tullis (1981). (1) Wet quartz; (2) dry quartz; (3) anothorite; (4) diopside; (5) wet olivine; (6) wet websterite; (7) dry olivine; (8) dry websterite.

where σ_x is horizontal stress in the x direction, z is depth and L is lithosphere thickness.

Application of stress to the lithosphere results in deformation. Material may deform elastically, by brittle failure and by plastic creep. The distribution of horizontal deviatoric tectonic stress with depth is controlled by the distribution of strength with depth. In the upper lithosphere the dominant non-elastic deformation mechanism is brittle failure. Here strength increases with confining pressure and therefore depth. In the deeper hotter lithosphere the dominant non-elastic deformation mechanism is plasticity. The plastic failure stress envelope decreases with depth due to increase in temperature. The lithospheric stress–depth relationship is shown schematically in figure 1*a*. A region of purely elastic deformation may exist between the regions of brittle and plastic deformation.

The plastic deformation mechanism experiences the greatest variability and therefore exerts the most control on the lithosphere stress–depth relationship. In figure 1*b* estimates of strain rate based on laboratory experiments are plotted against temperature for rocks and minerals relevant to lithospheric plastic deformation. The strain rate curves plotted in figure 1*b* assume a dislocation creep mechanism with a power law stress and temperature dependence for strain rate $\dot{\epsilon}$ given by:

$$\dot{\epsilon} = A \exp(-E/RT)\,\tau^n,$$

where T is temperature, τ deviatoric stress and E activation energy.

Strain rate increases strongly with increase in temperature for all materials. In the mantle plastic deformation is controlled by olivine, while in the crust plastic deformation is controlled by quartz (upper crust) or plagioclase (lower crust). At the same temperature the quartzo-feldspathic crust is substantially weaker than the olivine mantle.

For a geologically significant strain rate of $10^{-15}\,s^{-1}$ a critical temperature of 300 °C must be exceeded for plastic deformation of wet quartz while for olivine a

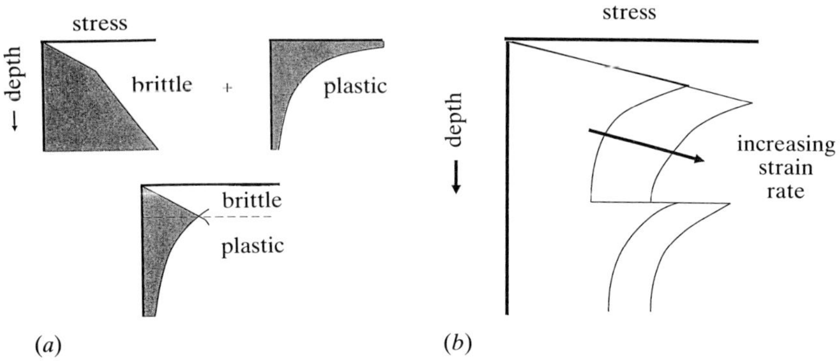

Figure 2. (*a*) Schematic representation of brittle and plastic strength with depth and their control on resultant stress with depth. (*b*) Schematic plot of stress versus depth for the constant strain rate model which assumes steady state deformation. Increasing strain rate increases stress within the plastic failure region and the force required to deform the lithosphere.

temperature of 600–700 °C must be exceeded (Kusznir & Park 1984, 1987). The strong temperature dependence of lithosphere rheology suggests that geothermal gradient will have a dominant control on the stress–depth relationship and therefore on the strength of lithosphere.

2. Constant strain rate and constant force models

(*a*) *Constant strain rate models*

In discussing the distribution of tectonic stress in the lithosphere it is often assumed that the whole of the lithosphere is extending or shortening at a constant strain rate (Chen & Molnar 1983). Within the upper lithosphere the stress–depth distribution follows the brittle failure envelope (figure 2*a*). In the region of plastic deformation beneath the brittle–plastic transition, the horizontal stress must sustain the constant horizontal strain rate with which the lithosphere is deforming and, because temperature increases with depth, the magnitude of the stress decreases with depth. The distribution of stress with depth for a constant strain rate is strongly controlled by strain rate (figure 2*b*). Increasing the horizontal strain rate requires larger values of stress within the plastic deformation region. This allows the region of brittle deformation to penetrate deeper.

The fundamental assumption of constant strain rate, however, is invalid before whole lithosphere failure. Furthermore the horizontal tectonic force required to sustain the deformation is often unrealistically high by several orders of magnitude and substantially greater than that available within the lithosphere.

(*b*) *Constant force models*

An alternative model for the prediction of the distribution of stress with depth is the constant force model (Kusznir 1982; Kusznir & Park 1984, 1987). Fundamental to this model is the assumption that the horizontal tectonic force F_x applied to the lithosphere is constant in time, i.e.

$$F_x = \int_0^L \sigma_x \, dz = \text{const.}$$

and
$$\int_0^L \dot{\sigma}_x \, \mathrm{d}z = 0.$$

Each infinitesimal component of the lithosphere is assumed to behave as a brittle–viscoelastic Maxwell material such that:

$$\epsilon_x = (1/E)(\sigma_x - \sigma_x^0) - (\nu/E)(\sigma_y - \sigma_y^0) - (\nu/E)(\sigma_z - \sigma_z^0) + \epsilon_x^{\mathrm{v}},$$
$$\epsilon_y = (1/E)(\sigma_y - \sigma_y^0) - (\nu/E)(\sigma_x - \sigma_x^0) - (\nu/E)(\sigma_z - \sigma_z^0) + \epsilon_y^{\mathrm{v}},$$
$$\epsilon_z = (1/E)(\sigma_z - \sigma_z^0) - (\nu/E)(\sigma_x - \sigma_x^0) - (\nu/E)(\sigma_y - \sigma_y^0) + \epsilon_z^{\mathrm{v}},$$

where ϵ is total strain, σ is stress, ϵ^{v} is viscous strain, σ^0 is initial stress for brittle failure, E is Young's modulus and ν is Poisson's ratio.

Furthermore each layer of the lithosphere is assumed to be welded together and suffers the same total horizontal strain so that:

$$\mathrm{d}\epsilon_x / \mathrm{d}z = 0.$$

The vertical stress σ_z arising from the application of the horizontal tectonic force is zero since the surface of the solid Earth is a free boundary, i.e. $\sigma_z = 0$.

The assumption of plane strain ($\epsilon_y = 0$) or plane stress ($\sigma_y = 0$) in the horizontal y direction perpendicular to the applied horizontal tectonic force F_x places additional constraints which permit the calculation of σ_x and σ_y in depth and time, t (Kusznir 1982):

$$\sigma_x = \int_0^t \left(\frac{1}{L} \int_0^L k\dot{\epsilon}_\nu \, \mathrm{d}z - k\dot{\epsilon}_\nu \right) \mathrm{d}t' - \frac{1}{L} \int_0^L \sigma_x^0 \, \mathrm{d}z + \sigma_x^0,$$

$$\sigma_y = \int_0^t \left(\nu\dot{\sigma}_x - E\frac{(2\sigma_y - \sigma_x)}{6\eta} \right) \mathrm{d}t' + \sigma_y^0 - \nu\sigma_x^0,$$

where k and $\dot{\epsilon}_\nu$ are defined below and η is apparent viscosity.

An alternative assumption to plane strain or plane stress is that both the x and y axes are unconstrained but with $\sigma_x = \sigma_y$ and $\epsilon_x = \epsilon_y$. Solutions for this assumption also exist for σ_x. The assumption that $\epsilon_x = \epsilon_y = 0$ has no physical meaning within the context of the constant force model. The values of k and $\dot{\epsilon}_\nu$ for these assumptions are:

plane strain in y axis:

$$k = E/(1-\nu^2), \quad \dot{\epsilon}_\nu = [\sigma_x(2-\nu) - \sigma_y(1-2\nu)]/6\eta, \quad \dot{\epsilon}_{\mathrm{T}} = (1+\nu)\,\alpha\Delta\dot{T};$$

plane stress in y axis:

$$k = E, \quad \dot{\epsilon}_\nu = \sigma_x/3\eta, \quad \dot{\epsilon}_{\mathrm{T}} = \alpha\Delta\dot{T};$$

unconstrained, y axis = x axis:

$$k = E/(1-\nu), \quad \dot{\epsilon}_\nu = \sigma_x/6\eta, \quad \dot{\epsilon}_{\mathrm{T}} = \alpha\Delta\dot{T};$$

ϵ_{T} is thermal strain and will be used later.

The application of the constant force model to continental lithosphere is shown in figure 3*a*. A compressive horizontal tectonic force of 10^{12} N m^{-1} has been applied to continental lithosphere with a heat flow of 60 mW m^{-2}. Crustal thickness is 35 km. Plastic deformation is assumed to be controlled by wet quartz in the upper crust, by dry quartz in the lower crust and by olivine in the mantle. Horizontal stress is plotted against depth for times of 10^3, 10^4, 10^5 and 10^6 years after the application of tectonic force. The initial elastic stress–depth distribution is shown by a dashed line.

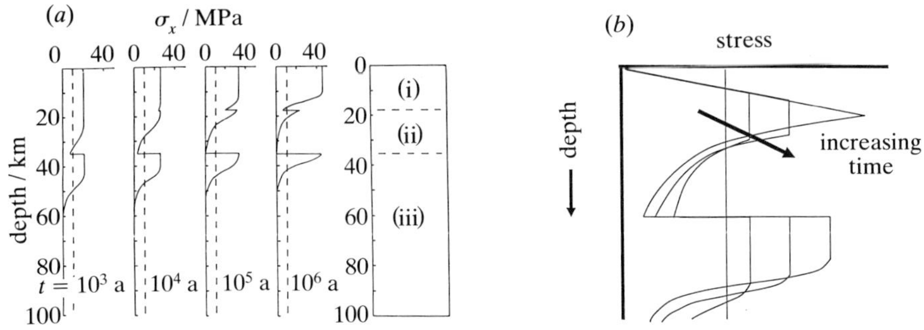

Figure 3. (*a*) Stress against depth at various times after the application of a tectonic force of 10^{12} N m^{-1} for the constant force model of continental lithosphere of average geothermal gradient. No brittle failure. $q = 60$ mW m^{-2}. (i) Upper crust 50% wet quartz; (ii) lower crust 10% dry quartz; (iii) mantle olivine. (*b*) Schematic stress–depth relationship for the constant force model. With increase in time, stress is concentrated and amplified into the upper lithosphere.

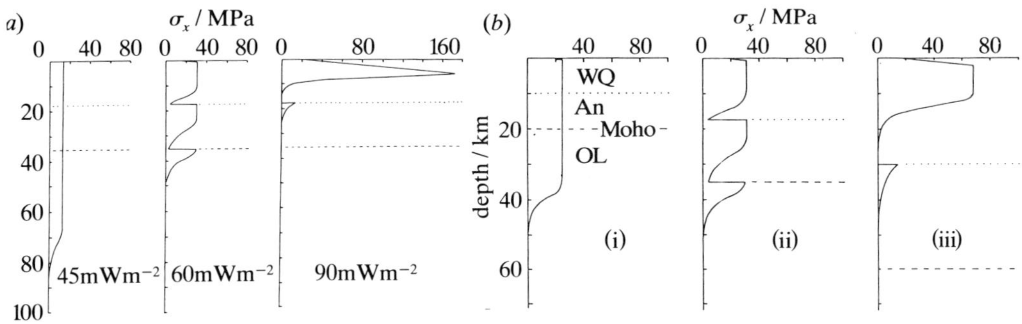

Figure 4. (*a*) Stress plotted against depth for continental lithosphere with cool, average and hot geothermal gradients at 1 Ma after the application of a tensile tectonic force of 10^{12} N m^{-1}. (*b*) Stress against depth for (i) thin (20 km), (ii) average (35 km) and (iii) thick (60 km) crust with average geothermal gradient at 1 Ma.

Plastic deformation in the lower lithosphere causes the decay of stress in the lower lithosphere. Because of force conservation the decayed stress is transferred upwards to increase stress in the upper lithosphere. This process has been called stress amplification by Kusznir & Bott (1977). As time progresses stress decay by creep occurs at higher levels within the mantle and the lower crust. By 1 Ma after the application of tectonic force, stress levels in the upper lithosphere have been amplified by a factor of four.

The behaviour of the constant force model with brittle failure also included is shown schematically in figure 3*b*. In contrast to the constant strain rate model, the constant force model predicts a transient stress distribution with depth and does not assume a constant strain rate with time.

3. Distribution of stress with depth in the continental lithosphere

The geothermal gradient has a dominant control on the lithosphere stress–depth distribution because of the strong temperature dependence of plastic deformation strain rates. In figure 4*a* horizontal stress is plotted against depth for a continental lithosphere with heat flows of 45, 60 and 90 mW m^{-2} subjected to a compressive

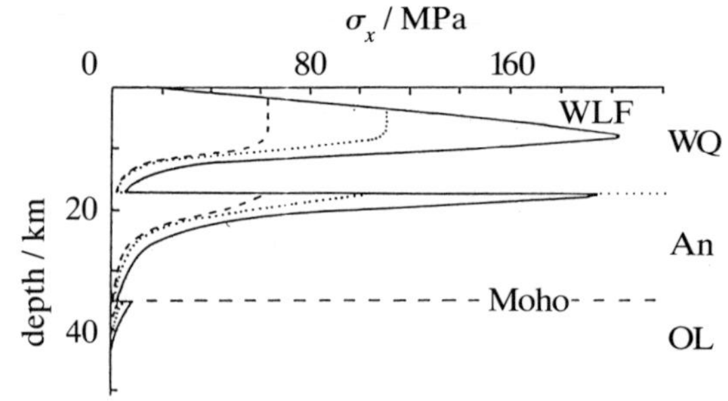

Figure 5. Stress plotted against depth for continental lithosphere for a range of magnitudes of extensional tectonic force after 1 Ma. Heat flow is 70 mW m^{-2}. Increase of the magnitude of the applied force results in WLF. ——, $F_x = 2.5 \times 10^{12}$ N m^{-1}; $\cdots\cdots$, $F_x = 1.5 \times 10^{12}$ N m^{-1}; -----, $F_x = 1.0 \times 10^{12}$ N m^{-1}.

tectonic force of 10^{12} N m^{-1}. Wet quartz, plagioclase and olivine rheologies are assumed for the upper crust, lower crust and mantle respectively. Crustal thickness is 35 km. For the cooler lithosphere with low heat flow, stress is carried to greater depth within the lithosphere (75 km for 45 mW m^{-2}). As heat flow increases, creep extends to a higher level in the mantle and lower crust until for 90 mW m^{-2} the stress is concentrated entirely into the upper crust. Substantial brittle failure occurs in the topmost lithosphere for the models with higher heat flows. For the 90 mW m^{-2} model, plastic creep and brittle failure cause stress amplification in the upper lithosphere by a factor of ten. Brittle failure is calculated using Griffith (1924) crack theory as modified by McClintock & Walsh (1962).

Because the quartzo-feldspathic rheology of the crust is much weaker at the same temperature than the olivine mantle, crustal thickness exerts a strong control on the stress–depth distribution of the continental lithosphere. This is plotted in figure 4*b* for crustal thicknesses of 20, 35 and 60 km for a fixed heat flow of 60 mW m^{-2}. For a thin crust, stress is carried to a much greater depth and less stress amplification occurs in the upper lithosphere because of the greater thickness of the stronger olivine rheology. As crustal thickness is increased to 60 km more stress relaxation by creep occurs in the lower crust and more stress is concentrated into the upper crust.

The level of stress amplification and the degree of upper lithosphere brittle failure are also controlled by the magnitude of the applied tectonic force.

Figure 5 shows the effect of increasing the applied extensional tectonic force from 1.0×10^{12} N m^{-1} to 2.5×10^{12} N m^{-1}. As tectonic force is increased there is an increasingly deeper penetration of brittle failure.

4. Whole lithosphere failure and the strength of the continental lithosphere

A typical stress–depth distribution, as predicted by the constant force model, is shown in figure 6*a*. The region of brittle failure in the upper lithosphere is separated from the region of plastic deformation in the lower lithosphere by a region of purely elastic deformation. Stress is maximum within this elastic core. As either the applied tectonic force or the geothermal gradient is increased, stress in the upper lithosphere increases permitting brittle failure to penetrate to a deeper level. Eventually the region of brittle failure and plastic deformation intersect (figure 6*a*) giving rise to whole lithosphere failure (WLF). Only after WLF has occurred can geologically significant strains in excess of a few percent occur (Kusznir & Park 1984).

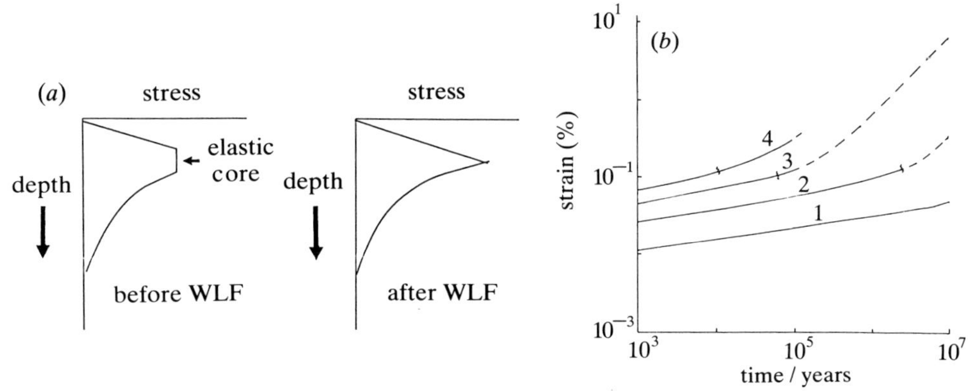

Figure 6. (*a*) Schematic representation of stress with depth for the constant force model showing an elastic core before WLF and its annihilation as WLF occurs. (*b*) Horizontal strain (%) plotted against time for a range of tensile force levels (in units of 10^{12} N m^{-1}) applied to continental lithosphere with heat flow of 70 mW m^{-2}.

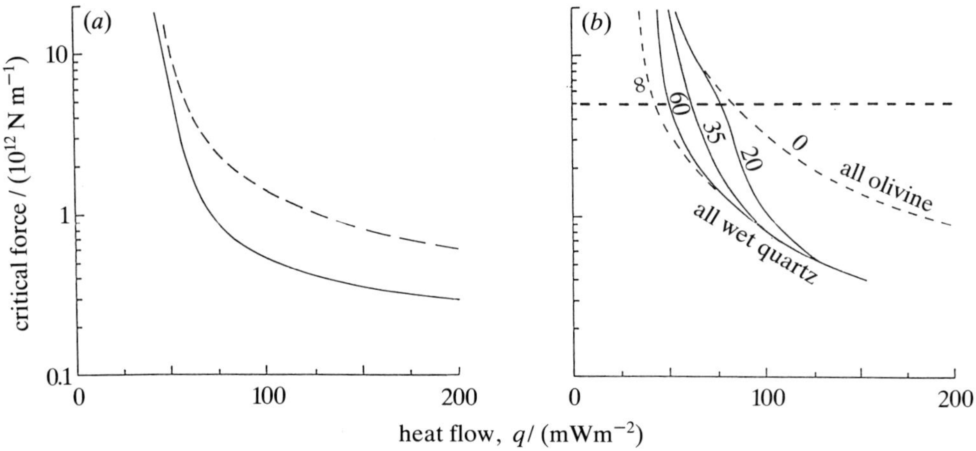

Figure 7. (*a*) Continental lithosphere extensional (——) and compressional (----) strength as a function of heat flow. (*b*) Lithosphere strength for a range of crustal thicknesses (km) with estimated maximum level of extensional tectonic force superimposed. ——, Quartz–olivine lithosphere; ----, end-member lithosphere.

Horizontal strain within the continental lithosphere for the constant force model, is plotted against time in figure 6*b* for different levels of applied extensional tectonic force for a heat flow of 70 mW m^{-2}. For large values of tectonic force WLF is achieved in a relatively short time (e.g. 10^4 years for 4×10^{12} N m^{-1}), while for smaller values a much longer time is required (greater than 10^7 years for 1×10^{12} N m^{-1}). After WLF the stress–depth relationship takes on a steady state and the horizontal strain rate becomes constant. WLF corresponding to a strain rate of 10^{-17} s^{-1} is only achieved after 1000 Ma following application of tectonic force. This implies that for slow constant strain rate models the steady state would never be achieved and that the constant strain rate model is invalid.

In figure 7*a* the force required to generate WLF within 1 Ma is shown as a function of surface heat flow for both extensional and compressional deformation of the lithosphere (Kusznir & Park 1984, 1987). Cool lithosphere in shield areas is inferred

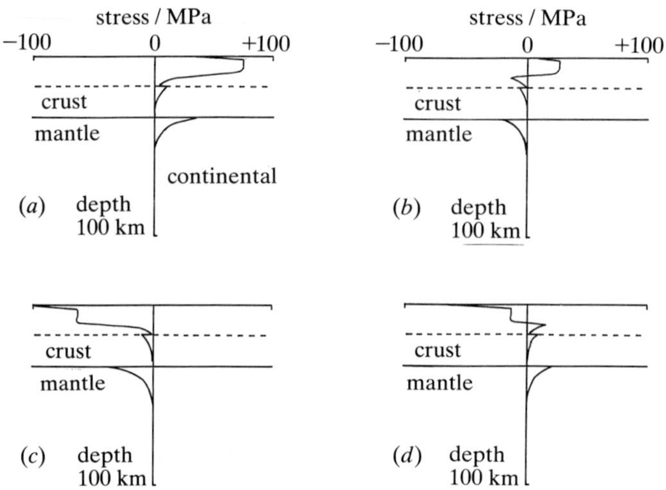

Figure 8. Stress against depth for continental lithosphere subjected to variation of tectonic force with time. See text for detailed explanation. (*b*) and (*d*) have zero net final applied force but non-zero stress.

to be substantially stronger than hot continental lithosphere. For a heat flow of 60 mW m^{-2}, corresponding to the average continental geothermal gradient, the extensional strength is *ca.* 2×10^{12} N m^{-1}. The lithosphere is stronger in compression than tension because of the stronger compressional brittle-failure envelope. In figure 7*b* the extensional strength of the continental lithosphere is shown as a function of heat flow for a range of crustal thicknesses; lithosphere with thick crust is substantially weaker than with thin crust because of the greater proportion of quartzo-feldspathic rheology.

Estimates of the magnitude of tectonic force arising from plate-boundary force and from isostatically compensated loads (Forsyth & Uyeda 1975; Chapple & Tullis 1977; Bott & Kusznir 1984) are:

Plate boundaries	Isostatically compensated loads
ridge push 2–3 $\times 10^{12}$ N m^{-1}	plateau uplift 0–4 $\times 10^{12}$ N m^{-1}
subduction suction 0–3 $\times 10^{12}$ N m^{-1}	continental margins 1–2 $\times 10^{12}$ N m^{-1}
subduction slab pull 0–5 $\times 10^{12}$ N m^{-1}	

The maximum available level of extensional tectonic force is superimposed on the extensional strength of the lithosphere in figure 7*b*. These strength estimates are comparable with the extensional strength for all but the coolest lithosphere in shield areas.

5. Time variation in tectonic force and the stress–depth relationship

Tectonic force within intraplate lithosphere is likely to vary with time as plate-boundary forces change. This has implications for the distribution of stress with depth in the lithosphere. In figure 8*a* the stress–depth distribution is shown 5 Ma after the application of an applied extensional tectonic force of 1×10^{12} N m^{-1} to continental lithosphere with a heat flow of 60 mW m^{-2}. A further incremental compressional force of -1×10^{12} N m^{-1} is applied after 5 Ma and the stress–depth distribution is left to evolve for a further 5 Ma. The result is shown in figure 8*b*. While

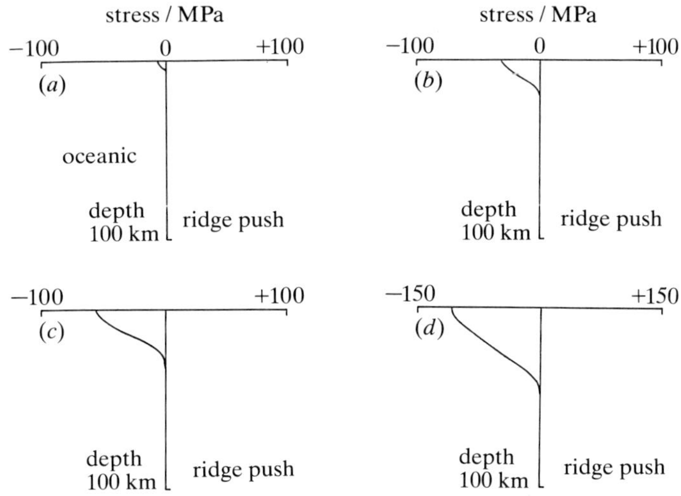

Figure 9. Horizontal stress against depth within the oceanic lithosphere arising from the ridge-push force as predicted by the constant force model. Stress is shown for ages of (*a*) 1 Ma, (*b*) 10 Ma, (*c*) 25 Ma and (*d*) 75 Ma.

the net tectonic force is zero, a complex distribution with depth of both compressive and tensile stress is seen.

The effect of applying a tensile tectonic force of 1×10^{12} N m^{-1} for 5 Ma, followed by an incremental compressive force of 2×10^{12} N m^{-1} for 5 Ma is shown in figure 8*c*. The effect of applying a further incremental tensile force of 1×10^{12} N m^{-1} for a third period of 5 Ma is shown in figure 8*d*. The final tectonic force is zero yet the stress–depth distribution shows both tension and compression. The stress-depth distribution clearly shows a 'memory' for the tectonic force that the lithosphere has experienced. The final models shown in figure 8*b*, *d* show very different stress–depth relationships but have identical zero net tectonic force.

6. Distribution of stress with depth in the oceanic lithosphere

Oceanic lithosphere is subjected to plate-boundary forces, to asthenospheric drag and to thermal cooling and contraction. These processes will result in a complex distribution of stress within the oceanic lithosphere.

If oceanic spreading ridges are assumed to be essentially lithostatic because of the low strength of the hot lithospheric material present there, then the ocean basins will be in absolute horizontal compression. The apparent compressive force responsible for this is called ridge-push force and may be estimated directly from anomalies in geoid height over oceanic spreading ridges (Haxby & Turcotte 1978; Dahlen 1981). The ridge-push force has been estimated to be of the order of 2.8×10^{12} N m^{-1} for lithosphere 80 Ma old. At the ridge crest the ridge-push force is zero. It has been assumed to increase linearly with age at a rate of 0.035×10^{12} N m^{-1} Ma^{-1}. The increasing ridge-push force is applied to a lithosphere which is cooling and becoming mechanically thicker as it ages.

Conservation of horizontal force within the lithosphere gives the equation:

$$\int_0^L \dot{\sigma}_x \, \mathrm{d}x = \dot{F}_{\mathrm{rp}},$$

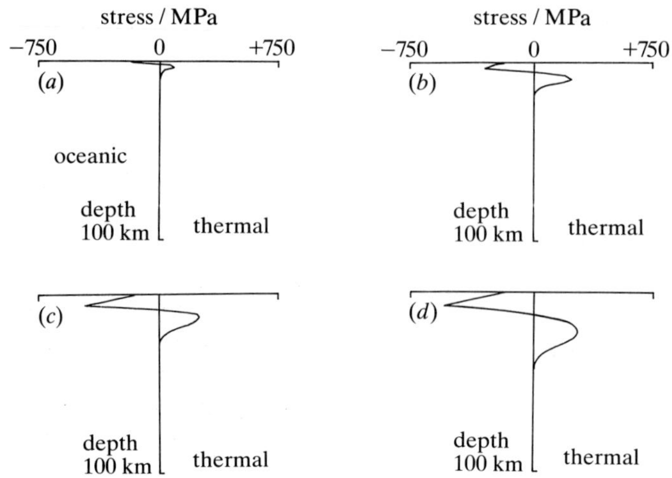

Figure 10. Horizontal thermal stresses within the oceanic lithosphere arising from the cooling of oceanic lithosphere as predicted by the constant force model. Stress is shown for ages of (*a*) 1 Ma, (*b*) 10 Ma, (*c*) 25 Ma and (*d*) 75 Ma.

where $\dot{F}_{\mathrm{rp}}$ is the rate of increase of ridge-push force. Again, the layers of the lithosphere are assumed to be welded together such that $\mathrm{d}\dot{\epsilon}_x/\mathrm{d}z = 0$.

The behaviour of horizontal stress σ_x with depth and time, assuming each infinitesimal component of the lithosphere behaves as a brittle–viscoelastic Maxwell material with power-law stress and temperature-dependent rheology can be shown to be given by:

$$\dot{\sigma}_x = \frac{k}{L}\int_0^L \left(\dot{\epsilon}_\nu + \dot{\epsilon}_{\mathrm{T}} + \frac{\dot{\sigma}_x^0}{k}\right)\mathrm{d}z - k\dot{\epsilon}_\nu - k\dot{\epsilon}_{\mathrm{T}} + \dot{\sigma}_x^0 + \frac{\dot{F}_{\mathrm{rp}}}{L},$$

where ϵ_{t} are thermal strains. k, ϵ_ν and ϵ_{T} have been defined earlier for both plane strain and plane stress in the y direction.

The horizontal stress within the oceanic lithosphere due to the ridge-push force is shown in figure 9 for oceanic lithosphere with ages of 1, 10, 25 and 75 Ma. An olivine rheology in the lithosphere and the thermal model of Parsons & Sclater (1977) have been assumed. As the lithosphere cools it also strengthens and stress is carried to a greater depth. Maximum compressive stress occurs at the surface with an almost linear decrease with depth to zero at about the 650 °C isotherm. For 75 Ma oceanic lithosphere the maximum compressive stress at the surface is about 120 MPa and is insufficient to generate brittle failure.

As the oceanic lithosphere cools as it ages and moves away from the oceanic spreading ridge, it contracts and will generate thermal stresses. The maximum cooling is of the order of 1300 °C and assuming a coefficient of thermal expansion α of 3.5×10^{-5} °C^{-1}, the maximum thermal strains are about 5%. For elastic deformation and a Young's modulus of 10^{11} N m^{-2} the maximum elastic thermal stress would be of the order of 5000 MPa. Clearly such large thermal stresses could not be sustained and would generate brittle and ductile deformation leading to decrease in thermal stress levels.

In figure 10 the result of applying the constant force model to the evolution of thermal stresses within oceanic lithosphere are shown at ocean lithosphere ages of 1, 10, 25 and 75 Ma assuming plane strain in the y direction. The assumption of plane stress in the y direction results in a slightly smaller, but otherwise similar, stress–depth distribution. The horizontal stress within the topmost lithosphere due to thermal contraction is compressive and is underlain by a region of horizontal tension. This is the opposite of a simple intuitive purely elastic model which would give tensile stresses in the upper lithosphere, where cooling has been greatest, above deeper compression. The reason for this fundamental difference is that both brittle and plastic deformation allow the release of the initially tensile stresses in the upper lithosphere created by cooling and contraction, so that when material below cools and contracts it takes the upper lithosphere into compression. The thermal stresses generate substantial brittle failure in compression in the upper lithosphere with some limited tensile brittle failure in the lower part of the mechanically competent lithosphere. After 75 Ma the maximum compressive horizontal stress is of the order of 600 MPa.

The combined effect of ridge-push and thermal-contraction stresses is shown in figure 11*a* for an oceanic lithosphere 75 Ma old. The stress–depth distribution is dominated by the thermal contraction but is biased towards compression due to the compressive ridge-push force. The horizontal stress in the upper lithosphere is again compressive above tension in the lower part of the plate.

Intraplate oceanic seismicity may be used to examine the state of stress of oceanic lithosphere (Wiens & Stein 1983, 1984; Bergman & Solomon 1985; Stein & Pelayo 1991). The distributions of epicentral depth and fault-plane solutions for oceanic intraplate seismicity are shown for comparison in figure 11*b* from work of Wiens & Stein (1983). Epicentral depths follow an isotherm between 600 and 700 °C consistent with the constant force thermo-rheological model and also with estimates of lithosphere thickness obtained from oceanic intraplate loading (Watts *et al.* 1980; Bodine *et al.* 1981). The fault-plane solutions show a tendency for the compressive events to lie in the upper part of the thermo-mechanically competent lithosphere while tensile events lie in the lower part. However, the observational data set of figure 11*b* is heavily weighted in favour of the Indian Ocean which may not be representative due to the anomalously high levels of tectonic force predicted there by Cloetingh & Wortel (1985, 1986).

7. Conclusions

1. Although the constant strain-rate model offers a simple method of calculating the horizontal stress–depth relationship within the lithosphere using brittle and plastic failure envelopes, it makes the assumption of steady-state deformation which is not valid in stable regions.

2. The constant force model incorporates elastic as well as brittle and viscous lithosphere behaviour and accommodates transient lithosphere deformation.

3. The constant force model predicts stress amplification within the upper lithosphere as a consequence of plastic deformation in the lower lithosphere. After stress amplification, levels of stress in the upper lithosphere are sufficient to generate brittle failure within the upper lithosphere.

4. The horizontal stress–depth distribution within the continental lithosphere arising from the application of a tectonic force is strongly dependent on geothermal gradient and crustal thickness.

5. For small horizontal strains, the lithosphere may show an elastic core between the brittle deformation of the topmost lithosphere and the plastic deformation of the middle and lower lithosphere. With increase in time, geothermal gradient or the level of the applied tectonic force, this elastic core may be annihilated as the brittle and

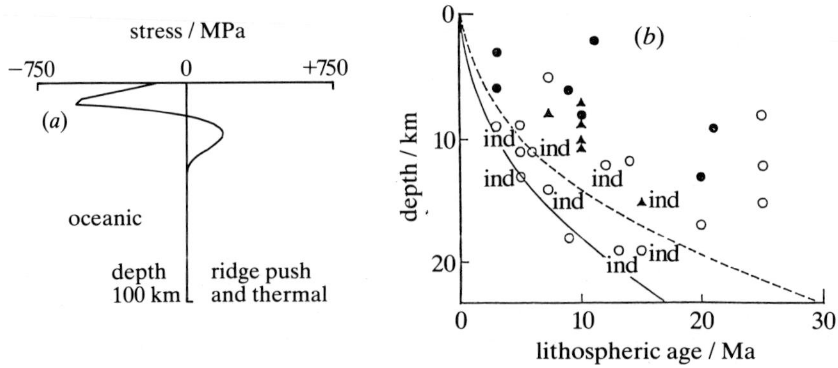

Figure 11. (*a*) Resultant stress arising from ridge push and thermal cooling for 75 Ma old oceanic lithosphere. (*b*) Depth and fault-plane solutions for oceanic intraplate seismicity against age (from Wiens & Stein 1983). Maximum depth of seismicity follows an isotherm between 600 and 700 °C. Type of faulting: ○, normal; ▲, strike–slip; ●, thrust.

plastic failure regions intersect and whole lithosphere failure occurs. Geologically significant strains can only occur after WLF has occurred.

6. The strength of the continental lithosphere, as defined by the level of tectonic force required to generate WLF, is strongly controlled by geothermal gradient and crustal thickness.

7. The levels of intraplate tectonic force arising from plate-boundary processes and isostatically compensated loads are comparable with the strength of the lithosphere calculated by the constant force model.

8. The constant force model predicts that the lithosphere stress–depth distribution has a memory of the history of tectonic force so that the stress–depth distribution will not be zero even if the tectonic force applied to the lithosphere is zero.

9. The compressive horizontal stresses within the oceanic lithosphere arising from ridge push are maximum at the surface of the oceanic lithosphere and decrease almost linearly to zero at a depth controlled by the 650 °C isotherm. The depth to which the ridge-push stress is carried by the lithosphere increases with lithosphere age.

10. Thermal horizontal stresses within oceanic lithosphere, as predicted by the thermo-rheological constant force model, are compressive in the upper lithosphere and tensile below. Thermal stresses are sufficient to generate substantial brittle failure.

11. The observed distribution of epicentral depths and fault-plane solutions for intraplate oceanic seismicity is consistent with the prediction of the constant force model which incorporates a combined ridge-push and thermal stress distribution.

I thank Martin Bott, Paul Caban, Graham Park and Andy Roberts for their inspiration, discussion, help and encouragement.

References

Ave Lallement, H. G. 1978 Experimental deformation of diopside and websterite. *Tectonophys.* **48**, 1–27.

Bergman, E. A. & Solomon, S. C. 1985 Earthquake source mechanisms from body-waveform inversion and intraplate tectonics in the northern Indian Ocean. *Phys. Earth planet. Int.* **40**, 1–23.

Bodine, J. H., Steckler, M. S. & Watts, A. B. 1981 Observations of flexure and the rheology of the oceanic lithosphere. *J. geophys. Res.* **86**, 3695–3707.

Bott, M. H. P. 1991 Sublithospheric loading and plate-boundary forces. *Phil. Trans. R. Soc. Lond.* A **337**, 83–93. (This volume.)

Bott, M. H. P. & Kusznir, N. J. 1984 Origins of tectonic stress in the lithosphere. *Tectonophys.* **105**, 1–14.

Chapple, W. M. & Tullis, T. E. 1977 Evaluation of the forces that drive the plates. *J. geophys. Res.* **82**, 1967–1984.

Chen, W.-P. & Molnar, P. 1983 Focal depths of intracontinental and intraplate earthquakes and their implications for the thermal and mechanical properties of the lithosphere. *J. geophys. Res.* **88**, 4183–4214.

Cloetingh, S. & Wortel, R. 1985 Regional stress field of the Indian plate. *Geophys. Res. Lett.* **12**, 77–80.

Cloetingh, S. & Wortel, R. 1986 Stress in the Indo-Australian plate. *Tectonophys.* **132**, 49–67.

Dahlen, F. A. 1981 Isostasy and the ambient state of stress in the oceanic lithosphere. *J. geophys. Res.* **86**, 7801–7807.

Fleitout, L. 1991 What are the sources of tectonic stresses? *Phil. Trans. R. Soc. Lond.* A **337**, 73–81. (This volume.)

Fleitout, L. & Froidevaux, C. 1982 Tectonics and topography for a lithosphere containing density heterogeneities. *Tectonics* **1**, 21–56.

Fleitout, L. & Froidevaux, C. 1983 Tectonic stresses in the lithosphere. *Tectonics* **2**, 315–324.

Forsyth, D. W. & Uyeda, S. 1975 On the relative importance of the driving forces of plate motion. *Geophys. Jl R. astr. Soc.* **43**, 163–200.

Goetze, C. 1978 The mechanisms of creep in olivine. *Phil. Trans. R. Soc. Lond.* A **288**, 99–119.

Griffith, A. A. 1924 Theory of rupture. In *Proc. 1st int. Congr. Applied Mechanics, Delft*, A **221**, 163–198.

Haxby, W. F. & Turcotte, D. L. 1978 On isostatic geoid anomalies. *J. geophys. Res.* **83**, 5473–5478.

Koch, P. S., Christie, J. M. & George, R. P. 1980 Flow law of 'wet' quartzite in the α-quartz field. *Eos, Wash.* **61**, 376.

Kohlstedt, D. L. & Goetze, C. 1974 Low-stress high-temperature creep in olivine single crystals. *J. geophys. Res.* **79**, 2045–2051.

Kusznir, N. J. 1982 Lithosphere response to externally and internally derived stresses: a viscoelastic stress guide with amplification. *Geophys. Jl R. astr. Soc.* **70**, 399–414.

Kusznir, N. J. & Bott, M. H. P. 1977 Stress concentration in the upper lithosphere caused by underlying viscoelastic creep. *Tectonophys.* **43**, 247–256.

Kusznir, N. J. & Park, R. G. 1984 Intraplate lithosphere deformation and the strength of the lithosphere. *Geophys. Jl R. astr. Soc.* **79**, 513–538.

Kusznir, N. J. & Park, R. G. 1987 The extensional strength of the continental lithosphere: its dependence on geothermal gradient, and crustal composition and thickness. In *Continental extensional tectonics* (ed. M. P. Coward, J. F. Dewey & P. L. Hancock), pp. 35–52. Geol. Soc. London Special Publ. 28.

McClintock, F. A. & Walsh, J. B. 1962 Friction on Griffith cracks under pressure. In *Proc. 4th U.S. Nat. Congr. Applied Mechanics*, pp. 1015–1021.

Parsons, B. & Sclater, J. G. 1977 An analysis of the variation of ocean floor bathymetry and heat flow with age. *J. geophys. Res.* **82**, 803–827.

Post, R. L. 1977 High temperature creep of Mt Burnet Dunite. *Tectonophys.* **42**, 75–110.

Shelton, G. & Tullis, J. 1981 Experimental flow laws for crustal rocks. *Eos, Wash.* **62**, 396.

Stein, S. & Pelayo, A. 1991 Seismological constraints on stress in the oceanic lithosphere. *Phil. Trans. R. Soc. Lond.* A **337**, 53–72. (This volume.)

Watts, A. B., Bodine, J. H. & Steckler, M. S. 1980 Observations of flexure and the state of stress in the oceanic lithosphere. *J. geophys. Res.* **85**, 6369–6376.

Wiens, D. A. & Stein, S. 1983 Age-dependence of oceanic intraplate seismicity and the implications for lithospheric evolution. *J. geophys. Res.* **88**, 6455–6468.

Wiens, D. A. & Stein, S. 1984 Intraplate seismicity and stress in young oceanic lithosphere. *J. geophys. Res.* **89**, 11442–11464.

Discussion

C. Vita-Finzi (*University College London, U.K.*). Professor Kusznir is testing the theory by reference to seismic data that assume that deeper events form part of a historical continuum in which the younger events are shallower. Is there some other way of testing the progressive change with time with reference to the oceans?

N. J. Kusznir. The observational data-set required to test the predicted oceanic stress–depth distribution consists of focal depths and fault-plane solutions as a function of lithospheric age. This data are very sparse and those which do exist are heavily weighted towards the Indian Ocean. However, I do not know of an alternative way of estimating the stress–depth distribution in the oceanic lithosphere.

R. B. Whitmarsh (*Institute of Oceanographic Sciences, Wormley, U.K.*). In the continental crust there is a strong/weak/strong sandwich with a weak lower crust. If a new rifted margin is started under tension presumably the lower crust is so weak it cannot transmit this tensional force. Would Professor Kusznir expect the upper crust, therefore, to separate away from the lower continental crust?

N. J. Kusznir. For continental lithosphere with an average geothermal gradient, horizontal tectonic force is carried within the upper crust and topmost mantle, with the lower crust acting as a low-stress–low-strength region. Although the lower crust may be weak, this does not in itself require that the upper crust and upper mantle deform by different amounts or that large-scale horizontal detachments exist within the lower crust.

P. England (*Oxford University, U.K.*). Professor Kusznir concluded that deforming the continental lithosphere at a constant strain rate is reasonable, provided the strain is of large magnitude. What is the time response of the system to changes in stress régime? The figures he showed seemed to depend on the interplay between two time constants. The first is a Maxwell time for the model lithosphere and the second is the rate at which externally applied forces can change. What is the second time constant and what is its size in relation to the Maxwell time constant?

N. J. Kusznir. The constant-force model of the lithosphere does not have a simply defined time-constant for stress decay and transfer. However, for most models the major part of stress variation occurs within 10^5–10^6 years.

Regarding the time-constant variations in tectonic force, I am sure that the amplitude spectrum of tectonic-force variation has finite amplitude at all frequencies. The geological record would suggest that very significant force fluctuations occur over periods of the order of 10^6–10^7 years. If this is so then fluctuations in tectonic force occur with a longer time-constant than the time taken for stress decay and transfer within the lithosphere.

M. H. P. Bott (*Durham University, U.K.*). What happens when a régime changes from tension to compression? What is the memory of the system? I wondered whether the timescale or the disappearance of the memory is of the order of 2 Ma.

N. J. Kusznir. The stress–depth relationship will be dependent on both the level of tensional and compressional force and the time elapsed, and not simply on the final level of tectonic force. The memory of tectonic force is 'encoded' in the stress–depth relationship and exists because of brittle or plastic failure; without brittle or elastic failure there would be no memory. The memory of tectonic force encoded in the stress distribution should not necessarily disappear with time. Only the application of large levels of tectonic force and whole lithospheric failure can destroy the lithosphere's memory of the stress–depth distribution.

R. Govers (*Utrecht University, The Netherlands*). How would the use of steady-state-flow laws for the variable strain rate flow modelling affect the time constants? I did not understand the physics behind the assumption that the vertical gradient of the horizontal strain rates is constant with depth due to the layers being welded together.

N. J. Kusznir. The creep rates used to calculate the stress–depth relationships in this paper assume steady-state deformation for wet and dry quartz, plagioclase and olivine. The mathematical formulation of the constant-force model could use transient creep; however, reliable transient-creep laws have not yet been determined experimentally.

The constant-force model assumes that total horizontal strain is constant with depth (equivalent to the layers of the lithosphere being welded together). Finite-element analysis confirms that this assumption is valid for distances greater than several lithosphere thicknesses away from the point of application of tectonic force (e.g. a plate boundary) for strains not exceeding a few percent.

P. England. Do the reversals of stress mentioned actually occur within a deforming zone because of the presence of tectonic forces which essentially are due to gradients in crustal thickness or gradients in buoyancy forces which do not just go away?

N. J. Kusznir. The prediction of the constant-force model, that reversals of stress with depth are generated by fluctuations in tectonic force, makes no assumptions about the origin of the horizontal tectonic force.

J. Cartwright (*Imperial College London, U.K.*). Has Professor Kusznir modelled the effects of lithosphere-scale discontinuities, such as shear zones inclined at 45°, which are inherently very weak?

N. J. Kusznir. No. The constant-force is a one-dimensional model assuming pure shear only.

S. Murrell (*University College London, U.K.*). The effect of pore pressure is very important both with respect to the level of stress and the depth of seismicity. This question should be addressed experimentally, not only in the laboratory but also in the field, for example by electrical conductivity measurements or by more detailed seismic measurements. I cannot understand how Professor Kusznir could believe that seismicity could occur in a plastically deforming medium. This has not been observed experimentally.

N. J. Kusznir. Pore pressure will have an important control on the level of stress and also on whether lithospheric material lies within the brittle or plastic deformation fields. The stress–depth distributions presented in this paper are calculated assuming zero pore pressure, corresponding to a maximum end-member of lithospheric brittle strength. Increasing pore pressure to hydrostatic (or above) will reduce the brittle-failure strength allowing brittle failure to penetrate to a greater depth within the lithosphere. The question remains as to what the pore pressure is within the middle and lower lithosphere.

Dynamics of the lithosphere and the intraplate stress field

By M. J. R. Wortel[1], M. J. N. Remkes[1], R. Govers[1], S. A. P. L. Cloetingh[2] and P. Th. Meijer[1]

[1] *Department of Geophysics, Institute of Earth Sciences, University of Utrecht, P.O. Box 80.021, 3508TA Utrecht, The Netherlands*

[2] *Department of Sedimentary Geology, Institute of Earth Sciences, Free University, De Boelelaan 1085, 1081HV Amsterdam, The Netherlands*

We outline the methodology of our numerical studies aimed at increasing the understanding of the relation between dynamics and stress field of the lithosphere with particular reference to oceanic lithosphere. The ridge-push force is modelled as a pressure gradient integrated over all contributing parts of the lithosphere. The slab-pull force is modelled as being dependent on the age of the subducting lithosphere. We parametrize the resistive forces and determine the unknown parameters by requiring the total torque of all forces acting on the plate to vanish. We illustrate the approach by the presentation and discussion of new modelling results for the Pacific plate.

1. Introduction

A variety of sources and processes is envisaged to contribute to the stress field in the lithosphere. Among these are dynamics of the lithosphere, heterogeneities in density structure and thickness of the crust and associated topography, bending processes and thermal stresses (see McNutt (1987) and Zoback *et al.* (1989) for recent reviews). In this paper we are concerned with the contribution arising from the dynamics of the lithosphere, that is the distribution of forces driving and resisting plate motion. Examples of studies in which the other contributions are addressed are Artyushkov (1973), Fleitout & Froidevaux (1982) and Froidevaux *et al.* (1988).

Studies of the dynamics of the lithosphere have followed two different approaches, both acknowledging that the dynamics of the Earth's lithosphere cannot be considered separately from the dynamics of the mantle: (*a*) one in which the plates are considered as principal units, and the interaction with the deeper mantle is parameterized in terms of forces or shear stresses acting along boundaries or the basal surface of the plates, and (*b*) the other in which emphasis is placed on the convecting mantle and the plates are considered to be the near-surface part of this system (Hager & O'Connell 1981; Forte & Peltier 1987; Vigny *et al.* 1991). Whereas the latter approach is appealing by virtue of its possible completeness and elegance, at present it does not allow as much detail of the stress field to be considered as does the former. Therefore, in this paper we focus on the former plate-wise approach for the dynamics of the lithosphere and use this approach as a basis for numerical calculations of the stress field in the lithosphere.

This approach requires a definition of the set of forces (referred to as 'plate-

Phil. Trans. R. Soc. Lond. A (1991) **337**, 111–126
Printed in Great Britain

tectonic forces') acting on lithospheric plates. Although pioneering studies of some of these forces, notably ridge push and slab pull, were made earlier (McKenzie 1969; Jacoby 1970), the study by Forsyth & Uyeda (1975) can be considered to mark the starting point of this plate force/torque approach (see also Chapple & Tullis 1977). They defined a set of forces acting on lithospheric plates, and assessed their relative importance. Also in search of the relevance of the various types of forces, Richardson *et al.* (1979) carried out an extensive series of numerical model calculations of the lithospheric stress field caused by plate-tectonic forces. The main difference between the studies by Richardson *et al.* (1979) and Forsyth & Uyeda (1975) and the approach outlined here (see also Wortel & Cloetingh 1981, 1983, 1985; Cloetingh & Wortel 1985, 1986) is that we consider some of the forces, in particular those driving the plates, to be well enough known to treat them as 'known' forces in the analysis. This is possible because the nature of the ridge push has been recognized to be that of an integrated pressure gradient (Lister 1975) and the dependence of this force and of slab pull (Vlaar & Wortel 1976) on lithospheric age has been established and because the age distribution of virtually all oceanic lithosphere is known (see Larson *et al.* 1985). By applying a balance of torques (of the forces acting on a plate), similar to that of Forsyth & Uyeda's (1975) study, quantitative information concerning the resistive forces can be obtained.

Because the forces acting on oceanic lithosphere are simpler (or better understood) than those acting on continental lithosphere, oceanic lithosphere provides a better opportunity to test whether plate-tectonic forces play a significant role in the stress field. Therefore, in this paper the emphasis will be on oceanic lithosphere. We outline basic aspects of our analysis of lithosphere dynamics and of the intraplate stress field and illustrate them by the presentation and discussion of new modelling results for the Pacific plate. Wherever this is appropriate, however, we draw on results obtained in earlier work.

2. Dynamics

(*a*) *General aspects*

The main aspects of the dynamic approach are illustrated in figure 1 (see also table 1), showing oceanic lithosphere with an accreting boundary (spreading ridge, on the right) and a subduction-zone boundary. With some rather straightforward modifications the analysis can be generalized to include continental lithosphere. In discussing the various forces we use the following definitions: (*a*) a force is specified per unit length of the boundary of the plate along which (or per unit area over which) it acts on the plate; the total force of each type acting on a particular plate is obtained by integration over the appropriate boundary or area, and (*b*) each force (say $\boldsymbol{F}_{\mathrm{a}}$) is factorized into a product of a scalar quantity (F_{a}) representing the magnitude of the force per unit length of a boundary (for boundary forces) or per unit area (for surface forces) and a unit vector ($\hat{\boldsymbol{e}}_{\mathrm{a}}$) giving the orientation of the force: $\boldsymbol{F}_{\mathrm{a}} = F_{\mathrm{a}}\hat{\boldsymbol{e}}_{\mathrm{a}}$.

The forces which drive the plates are the ridge push $\boldsymbol{F}_{\mathrm{rp}}$ and the slab pull $\boldsymbol{F}_{\mathrm{sp}}$. The ridge-push force is not a boundary force acting at the ridge axis but a pressure gradient integrated over the area of the plate (Lister 1975). This pressure gradient results from the cooling and densification of the lithosphere. Therefore, contributions to the ridge push come from all parts of the oceanic lithosphere which are subject to continued cooling, that is all oceanic lithosphere younger than approximately 90 Ma (which is taken as the cut-off age).

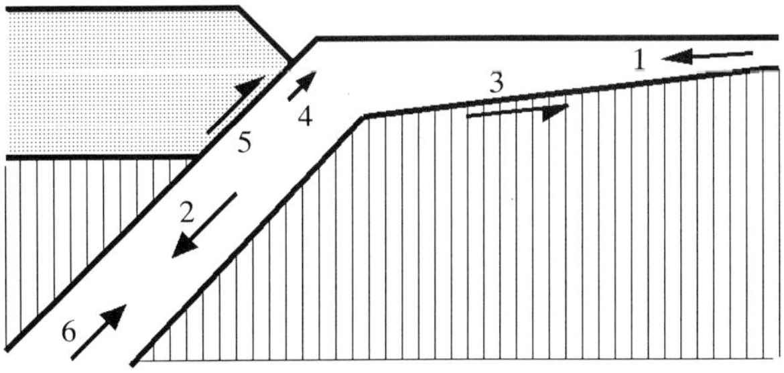

Figure 1. Schematic cross section of oceanic lithosphere, including ridge and trench, with an indication of forces used in the modelling: 1, ridge push F_{rp}; 2, slab pull F_{sp}; 3, drag force at the base of the lithosphere F_{dr} (alternatively called shear stress σ_b); 4, compositional buoyancy force F_{cb}; 5, resistance at plate contact F_{pc}; 6, deep resistance to slab penetration F_{sr}. Not shown are F_{up} (7), the force acting (at the plate contact) on the upper (overriding) plate, and F_{tf} (8), the resistive force acting along transform boundaries.

Table 1. *Force definition (see also figure 1)*

no.	force	symbol	no.	force	symbol
1	ridge push	F_{rp}	5	plate contact resistance	F_{pc}
2	slab pull	F_{sp}	6	deep resistance	F_{sr}
3	drag force	F_{dr}	7	upper plate resistance	F_{up}
4	compositional buoyancy	F_{cb}	8	transform fault resistance	F_{tf}

The ridge push, for a section (of unit width) of lithosphere from spreading ridge to age t, is calculated after Richter & McKenzie (1978),

$$F_{rp} = gh[\rho_m - \rho_w][\tfrac{1}{3}L + \tfrac{1}{2}h], \tag{1}$$

where g is the gravitational acceleration, h is the difference in elevation between the ridge crest and the ocean floor, L is the lithospheric thickness and ρ_w and ρ_m are the densities of seawater and of the sublithospheric mantle, respectively. Both L and h increase with age t according to a square-root-of-age relation, at least for $t < 70$ Ma (Crough 1975). This results in:

$$F_{rp} = c_{rp}\,t, \tag{2}$$

where c_{rp} is a constant. In our modelling the total ridge-push force is distributed over the area of the plate, on the basis of the age distribution and the implicit spreading history.

The other driving force, the slab pull F_{sp}, is the downdip component of the gravitational body force acting on the dense subducted slab (McKenzie 1969; Jacoby 1970). Per unit length in the direction of the trench this can be expressed as:

$$F_{sp} = g\sin\phi \int_0^L \int_0^{S_{sz}} \Delta\rho \, \mathrm{d}x \, \mathrm{d}z. \tag{3}$$

Here, ϕ is the dip and L the thickness of the downgoing slab, and $\Delta\rho$ the density contrast between slab and surrounding mantle. The x- and z-directions are downdip and perpendicular to the upper (dipping) interface between the slab and the upper mantle, respectively. The downdip integration is limited to the end of the seismic zone (S_{sz}), which is reasonable in view of the drastic change in constitution (rheology) and subsequent deformation beyond this point (Wortel & Vlaar 1988). Using McKenzie's (1969) thermal model we find

$$F_{sp} = (4g\alpha\rho_m T_m L^3/\kappa\pi^4)v_c \sin\phi[1-\exp(-\pi^2\kappa S_{sz}/v_c L^2)], \tag{4}$$

where v_c is the rate of plate convergence. The parameters α, ρ_m, T_m and κ are the coefficient of thermal expansion, the density and the temperature of the asthenosphere, and the thermal diffusivity of the asthenosphere/lithosphere respectively. For the present purpose, all these parameters may be taken to be constant. Using the age dependence of L (see above), $\kappa = 0.9\times10^{-6}\ \mathrm{m^2\,s^{-1}}$ and $S_{sz} = 0.12\ v_c t$ (see Wortel & Vlaar 1988) we obtain

$$[1-\exp(-\pi^2\kappa S_{sz}/v_c L^2)] = 0.29.$$

Thus the factor between brackets in (4) depends neither on v_c nor on the age of the lithosphere. Putting $v_c \sin\phi = v_z$, it is found that

$$F_{sp} \propto v_z L^3. \tag{5a}$$

Taking into account the age-dependence of L gives

$$F_{sp} = c_{sp}\, v_z\, t^{\frac{3}{2}}, \tag{5b}$$

where c_{sp} is a constant. On the basis of their established age-dependence the driving forces $\boldsymbol{F}_{rp}$ and $\boldsymbol{F}_{sp}$ can be calculated. Information on the age distribution of oceanic lithosphere is taken from Larson *et al.* (1985).

The resistive forces shown in figure 1 (see also table 1) are the drag force $\boldsymbol{F}_{dr}$ acting at the base of the lithosphere (alternatively referred to as shear stress σ_b at the base of the plate) and three resistive forces acting on the downgoing slab in the trench region and in the upper-mantle part of the subduction zone. These three resistive forces, all of which are expressed as boundary forces per unit length in the direction of the strike of the subduction zone (trench axis), are: the compositional buoyancy force $\boldsymbol{F}_{cb}$ (acting in the up-dip direction of the subducting slab) and resistive shearing forces $\boldsymbol{F}_{pc}$ (representing the integrated value of the shear stress σ_{pc} along the plate contact with the overriding plate) and $\boldsymbol{F}_{sr}$ (the integrated value of σ_{sr}, the resistance to slab penetration in the upper mantle, taken to act upwards parallel to the dip of the slab). $\boldsymbol{F}_{cb}$ is associated with the petrological stratification of the oceanic crust and upper mantle created during the spreading process (Vlaar & Wortel 1976; Oxburgh & Parmentier 1977; England & Wortel 1980); both the basaltic crust and the depleted uppermost mantle are less dense than the adjacent upper mantle would be at an equivalent temperature and pressure. This density structure contributes a gravitationally stable component to the lithosphere/asthenosphere system. Upon subduction this is felt as a force ($\boldsymbol{F}_{cb}$) which counteracts the gravitationally unstable component resulting from the cooling and densification of the lithosphere which is represented by $\boldsymbol{F}_{sp}$. As noted by England & Wortel (1980) it is probable that the compositional buoyancy of the descending lithosphere is removed by phase changes at less than 100 km depth. Consequently, we estimate the magnitude of this force by integrating the (compositional) density difference of the dipping oceanic lithosphere down to a depth of 100 km (value given below). Not displayed in the cross section of figure 1, but included in the modelling, are the resistive force $\boldsymbol{F}_{tf}$ along transform faults, and the force $\boldsymbol{F}_{up}$ acting on the upper plate at the plate contact (counterpart of $\boldsymbol{F}_{pc}$ acting on subducting plate). In specific cases, different from the entirely oceanic plate considered in figure 1, some other forces (not indicated in figure 1) may be relevant, like drag beneath the continents and the forces associated with continental collision (see Cloetingh & Wortel 1985, 1986).

To ensure mechanical equilibrium, the net torque (with respect to the Earth's centre) exerted by the forces on a plate is required to vanish:

$$\sum_{N} \int_{P} \boldsymbol{r} \times \boldsymbol{F} \, \mathrm{d}p = 0, \tag{6}$$

where the sum is over the total set of N forces and the integration is over the boundaries or the area of the plate P, depending on the nature of the force involved (boundary force or surface force). Of the forces considered above, the drag force at the base of the lithosphere ($\boldsymbol{F}_{\mathrm{dr}}$, or shear stress σ_{b}) and the ridge push ($\boldsymbol{F}_{\mathrm{rp}}$) are distributed over the area of the plate, whereas the other forces (see table 1, figure 1) are all modelled as boundary forces, either along the convergent (subduction zone) boundaries or (in case of $\boldsymbol{F}_{\mathrm{tf}}$) along transform fault boundaries. In the case of subduction-zone boundaries the trench axis is adopted as the geometrical boundary of the plate.

Solomon & Sleep (1974) applied a balance of torques to the combined set of all plates. In their study, which was directed at assessing the absolute motions of the plates, this was a useful formulation. As we shall proceed from dynamics and motion towards stress calculation we are interested in the entire set of forces acting on a plate, including forces distributed symmetrically along plate boundaries which do not contribute to a globally integrated torque but do contribute to the torque integrated for an individual plate.

(*b*) *Application and results*

We illustrate the procedure for the case of the Pacific plate. Figure 2 gives the age distribution (after Larson *et al.* 1985) used in the calculation of the slab pull and ridge push. Additional age data for the relatively young (Miocene–Oligocene) region southwest of the Marianas were incorporated after Weissel & Anderson (1978).

Several absolute motion models (Minster & Jordan 1978 (AM1-2); Chase 1978; Gripp & Gordon 1990 (HS2-NUVEL1)) have been used in the analysis. The difference between results obtained with different absolute motion models appeared to be insignificant. Values displayed or given below are those for the HS2-NUVEL1 model (Gripp & Gordon 1990).

Relative-motion models are generally derived from a data set comprising spreading rates at oceanic spreading centres, transform fault azimuths and slip vectors of earthquakes at the plate contact in subduction zones (see Demets *et al.* (1990) for a recent application and results). For the relative motion between the Pacific plate and neighbouring plates the various relative motion models are similar. For our modelling we used the relative motion data according to Scotese *et al.* (1988).

Whereas it is (relatively) clear, in particular with the nature of the ridge push as given above, where the driving forces act upon the lithosphere, it is considerably less evident where the various resistive forces (except $\boldsymbol{F}_{\mathrm{cb}}$ and $\boldsymbol{F}_{\mathrm{tf}}$) are exerted. Our analysis yields a significant trade-off for resistive-force magnitudes.

To reduce the number of unknowns we choose to adopt values for some resistive forces. On the basis of model results obtained by vanden Beukel (1990) we choose $F_{\mathrm{cb}} = 3.0 \times 10^{12}\ \mathrm{N\ m^{-1}}$, and for the force associated with transform faults we take

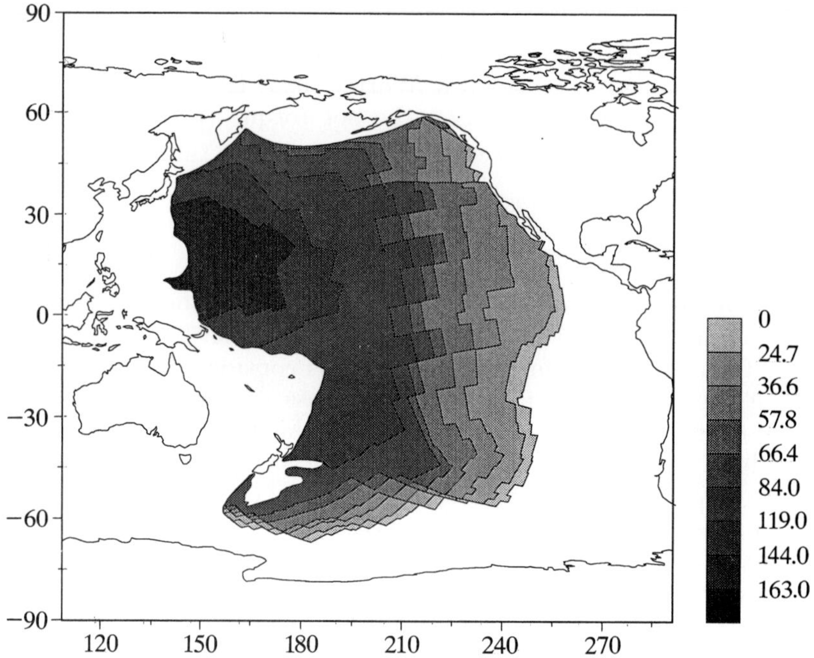

Figure 2. Age distribution of Pacific plate (after Larson *et al.* 1985). Values in legend are in millions of years.

$F_{tf} = 0.5 \times 10^{12}$ N m^{-1}. The latter value is in agreement with shear-stress values inferred for the San Andreas fault system (Zoback *et al.* 1987); we note, however, that the analysis appears to be quite insensitive to reasonable variations in the assumed value for F_{tf}. The remaining unknown resistive forces (or shear stresses) are $\boldsymbol{F}_{pc}$, $\boldsymbol{F}_{sr}$ and $\boldsymbol{F}_{dr}$ (or σ_b). The directions of these forces follow from the adopted relative and absolute motion models; and, theoretically, the three magnitudes F_{pc}, F_{sr} and σ_b (each of which is assumed uniform for the plate) can be determined by solving vector equation (6). The resulting solution, however, shows a negative value for the magnitude F_{pc} which corresponds to a driving force. For a plate contact force this is considered to be physically unrealistic. Therefore, negative values of F_{pc} are not accepted. By not accepting negative values of F_{pc}, however, we leave out the solution with a zero net torque. Hence, any other solution will show a net torque residue. The minimum value for the torque residue was found for $F_{sr} = 2.0 \times 10^{13}$ N m^{-1}. The corresponding trade-off relation between F_{pc} and the shear stress at the base of the lithosphere σ_b is shown in figure 3.

The torque residues have been normalized by the magnitude of the torque of the slab pull of the 'old' subduction zones (i.e. zones consuming lithosphere older than 90 Ma). For a given absolute motion model the trade-off relation between σ_b and F_{pc} is entirely a function of the geometry of the plate (surface area) and the distribution of subduction zones along its plate boundaries.

The torque residue appears to be rather insensitive to variations in σ_b for values less than *ca.* 0.3 MPa (3 bar). The minimum in the residues corresponds to a value for F_{pc} equal to zero. This is still physically unrealistic, and for the subsequent stress calculations we avoid such a non-positive value (i.e. a non-resistive plate contact

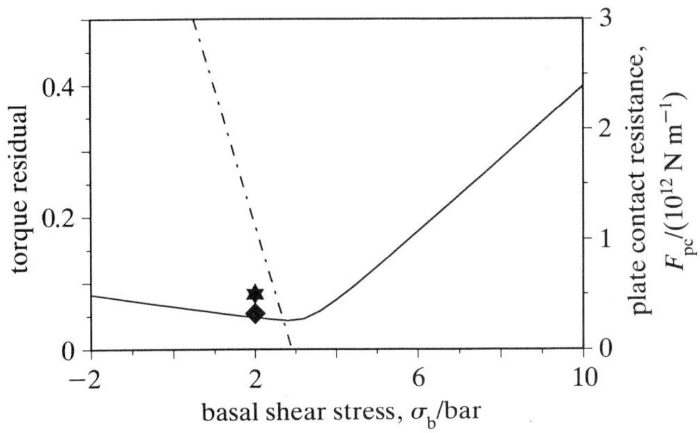

Figure 3. Trade-off in resistive forces. ––––––, Trade-off between shear stress at the base of the lithosphere σ_b and the resistance at the plate contact F_{pc} (vertical axis on the right). 10 bar = 1 MPa. ——, Residual of net integrated torque normalized by the magnitude of the torque of the slab pull of the 'old' slabs (vertical axis on the left). The asterisk indicates the values of the combination of σ_b and F_{pc} used in the stress calculation; the black square symbol indicates the corresponding residual torque. Deep resistance: $F_{sr} = 2 \times 10^{13}$ N m^{-1}.

Table 2. *Torque results*[a]

force		T_x	T_y	T_z	magnitude	lon. (°E)	lat. (°N)
ridge push	F_{rp}	1.836	2.011	−14.404	14.659	47.61	−79.29
slab pull (old)	F_{spo}	−3.015	57.462	−93.751	110.001	92.99	−58.45
slab pull (young)	F_{spy}	−0.096	12.756	−1.496	12.844	90.42	−6.69
drag force	F_{dr}	1.860	−5.961	10.594	12.297	−72.67	59.48
buoyancy	F_{cb}	0.599	−16.835	15.033	22.578	−87.95	41.74
plate contact (old)	F_{pco}	0.540	−0.677	2.152	2.320	−51.43	68.08
plate contact (young)	F_{pcy}	1.081	−0.757	1.511	2.006	−34.98	48.86
deep resistance	F_{sr}	2.441	−46.523	75.903	89.060	−86.99	58.45
upper plate	F_{up}	−0.466	0.106	0.874	0.996	167.33	61.34
transform fault	F_{tf}	0.664	−0.119	1.695	1.824	−10.17	68.28

[a] Torques are specified in units of 10^{25} N m, using a coordinate system with the origin in the Earth's centre and the x-axis through (0° N, 0° E), the y-axis through (0° N, 90° E) and the z-axis through the North Pole.

resistance) in our modelling. Therefore, we choose the combination of σ_b and F_{pc} indicated by the asterisk (see also §3).

The numerical values of the resulting torque vectors are given in table 2. Figure 4 gives a schematic display. The ridge-push torque is about an order of magnitude smaller than the slab-pull torque. This partially results from the age distribution in the northern and southern parts of the Pacific plate (E–W isochrons) which causes an internal compensation of the ridge-push torque.

Of particular interest is the question of compensation of the slab-pull forces by resistive forces acting in the subduction-zone area. As Solomon & Sleep (1974) and Chapple & Tullis (1977) realized, net subduction forces exerted on a plate from a long convergent plate-boundary may give rise to intraplate stresses of considerable magnitude. Indeed, our way of modelling the dynamics of the plate results in net forces acting along subduction zones. They are responsible for the tensional stresses in the northwestern part of the Pacific plate (§3, figure 5).

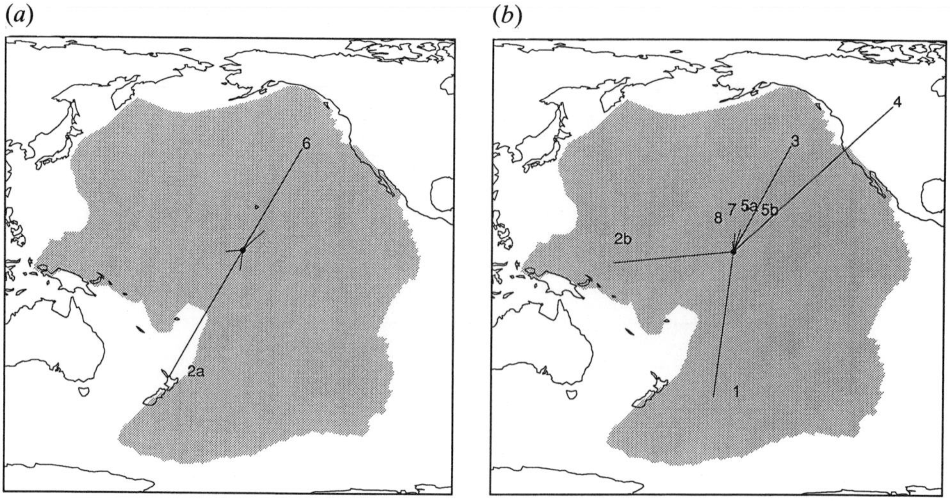

Figure 4. Schematic representation of torques: 2a and 2b indicate torque of slab pull for subduction zones for which the age of the descending lithosphere is older and younger than 90 Ma, respectively. A similar division (5*a* and 5*b*) is made for $\boldsymbol{F}_{pc}$, the plate contact resistance. Other numbers correspond to those given in figure 1 (see also table 1). (*a*) All torques (not all are visible); (*b*) enlarged version of (*a*), but without torques of 'old' slab pull (2a) and deep resistance (6).

Adding the torques $\boldsymbol{T} = \int_{L_{conv}} \boldsymbol{r} \times \boldsymbol{F} \, dL$ of all forces acting in or along the convergent plate boundaries (L_{conv}) of the Pacific plate gives

$$|\boldsymbol{T}_{sp} + \boldsymbol{T}_{up} + \boldsymbol{T}_{pc} + \boldsymbol{T}_{cb} + \boldsymbol{T}_{sr}|/|\boldsymbol{T}_{sp}| = 0.05.$$

Thus upon torque integration over the convergent boundaries the compensation is nearly complete.

Finally, in spite of the insensitivity of the torque residue to variations in σ_b, our results do show some preference for positive values of σ_b, which in our analysis correspond to a resistive shear-stress at the base of the lithosphere.

3. Stress

(*a*) *General aspects*

For a rheological model of the lithosphere the stress field in the lithosphere associated with the plate-tectonic forces can be calculated numerically. This field, which is called the regional stress field, represents the non-lithostatic state of stress in the lithosphere. (It is often referred to as the deviatoric stress field, although the stress tensor is not necessarily deviatoric.) Because the stress values are to be compared with the strength distribution in the lithosphere, which is expressed in terms of differential stress, the lithostatic part of the stress field (depth-dependent pressure) can be omitted. Finite-element codes are used to carry out the calculations. The mesh for the plate consists of an assembly of triangular elements. The plates are modelled (with a plane stress approximation) as being elastic with a reference thickness of 100 km, Young's modulus $E = 7 \times 10^{10}$ N m^{-2} and Poisson's ratio $\nu = 0.25$. No physical significance should be attached to this chosen reference thickness. From model results for this rheology the stresses in a plate with a non-uniform thickness or a more complex rheology (see Goetze & Evans 1979) can be derived. As an example we mention the study by Stein *et al.* (1987) in which stresses calculated

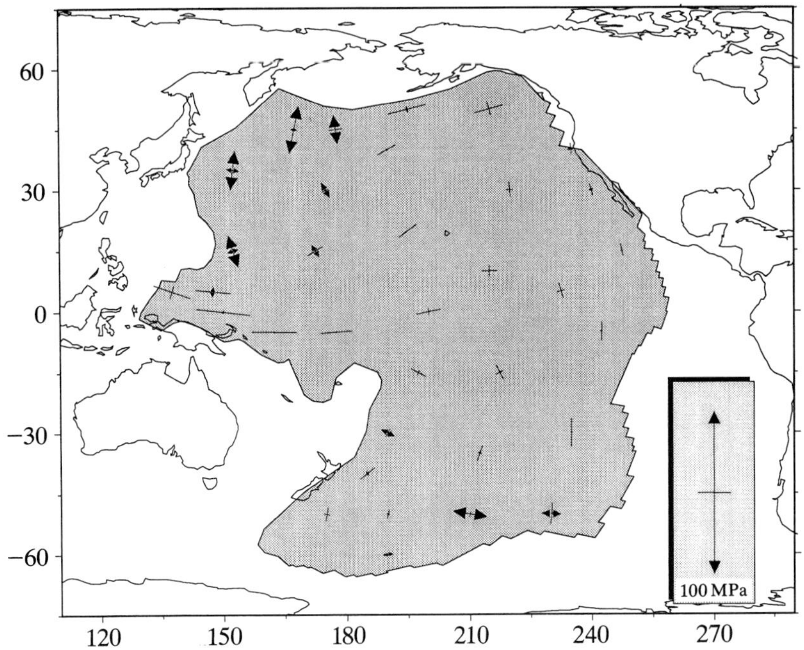

Figure 5. Calculated stress field of the Pacific plate. Plotted are the principal horizontal non-lithostatic stresses averaged over a uniform elastic plate with a reference thickness of 100 km. Principal axes with arrow-heads indicate tension, without arrow-heads compression. Scale is given in the lower-right corner.

for the Indo-Australian plate, on the basis of an assumed uniform plate thickness (100 km), were rescaled to stresses in the plate with an age-dependent thickness.

(*b*) *Stress field of the Pacific plate*

In principle it would be interesting to investigate the effect which varying the combination of F_{pc} and σ_b (according to figure 3) would have on the calculated stress field. Any non-zero torque, however, introduces reaction forces in the numerical procedure resulting in non-realistic features in the stress field. The magnitude of these forces increases with increasing magnitude of the torque residue. Combining the need for the torque residue to be as small as possible and the choice (see §2) that F_{pc} should be non-zero and positive led us to select the combination of values of F_{pc} and σ_b indicated by the asterisk in figure 3. The corresponding torque residue (square symbol in figure 3) is very near the minimum value.

The computational results for the intraplate stress field are shown in figure 5. The level of intraplate stress is moderate, typically in the order of 30 MPa (300 bar) or less (as displayed for a reference thickness of the plate).

4. Tests

(*a*) *Dynamics and plate motion*

We can test the force modelling by the procedure proposed by Gordon *et al.* (1978). They proposed that the absolute motion of a plate is governed by the integrated torque of the driving forces acting on the plate, slab pull and ridge push. In other words, the torque of the driving forces parallels the rotation vector describing

Figure 6

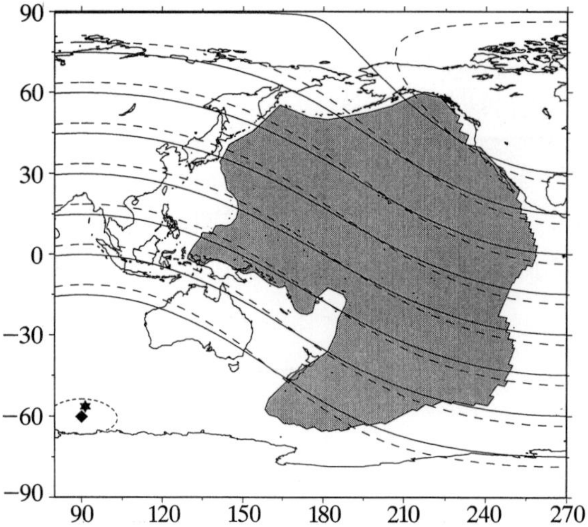

Figure 7

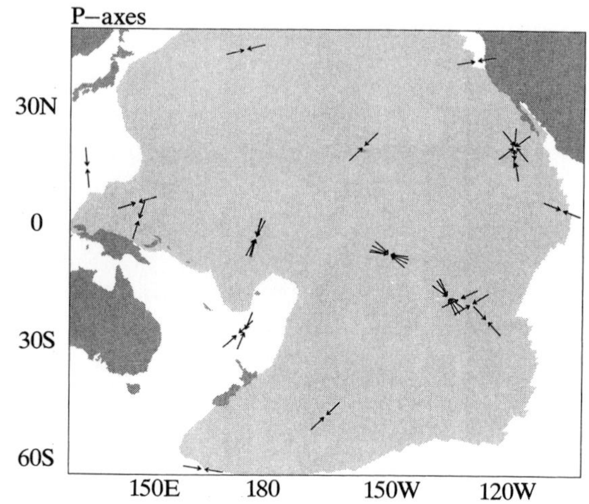

Figure 6. Relation between driving forces and absolute motion for the Pacific plate. Black square symbol: pole of absolute motion vector according to HS2-NUVEL1 (Gripp & Gordon 1990; see also DeMets *et al.* 1990). Solid curves are corresponding small circles. Asterisk: pole of integrated torque vector of driving forces (slab pull and ridge push). Dashed curves are corresponding small circles.

Figure 7. P-axes of focal mechanisms of intraplate earthquakes in the Pacific (World Stress Map database; Zoback *et al.* 1989).

absolute plate-motion. In their original approach Gordon *et al.* (1978) made the assumption that slab pull and ridge push are constant per unit width and act perpendicularly to the relevant segments of the trench and the ridge respectively. If now we quantify the slab pull and the ridge push as forces dependent on lithospheric age and consider the ridge push no longer as a line force but as a distributed force (see above) we can apply the same test to the Pacific plate. Figure 6 shows the pole of the torque of $\boldsymbol{F}_{sp}$ and $\boldsymbol{F}_{rp}$ integrated over the plate. The predicted rotation pole falls within the error ellipse for the absolute motion model HS2-NUVEL1 (Gripp & Gordon 1990) which is based on Minster & Jordan's (1978) hot spot data and the NUVEL1 relative motion model (DeMets *et al.* 1990). This is in excellent agreement with the proposition of Gordon *et al.* (1978).

(*b*) *Stress field: orientation and magnitude*

(i) *Orientation*

The orientations of the principal stress axes as shown in figure 5 can be compared with observational data. The World Stress Map project was aimed at providing a data base for this purpose. For the Pacific plate, however, the available data-set is very limited in both quantity and quality (see Zoback *et al.* 1989). Nevertheless reasonable agreement between the calculated field (figure 5) and the focal mechanism data (figure 7), notably the rotation of stress axes in the central part of the Pacific plate and E–W oriented compression in the northwestern part, can be observed. The low level of seismic activity seems to be in qualitative agreement with the low level of stress inferred from our modelling.

For oceanic intraplate stress fields the best opportunity to employ stress information from seismicity to test our stress modelling is provided by the Indo-Australian plate. There are two reasons: (1) the level of seismic activity is high by

Figure 8

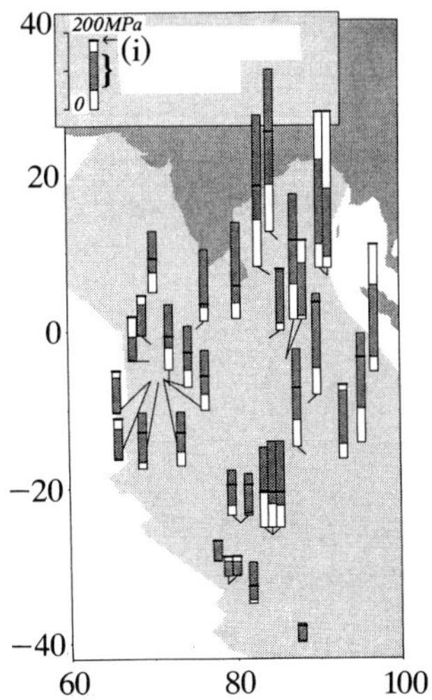

Figure 9

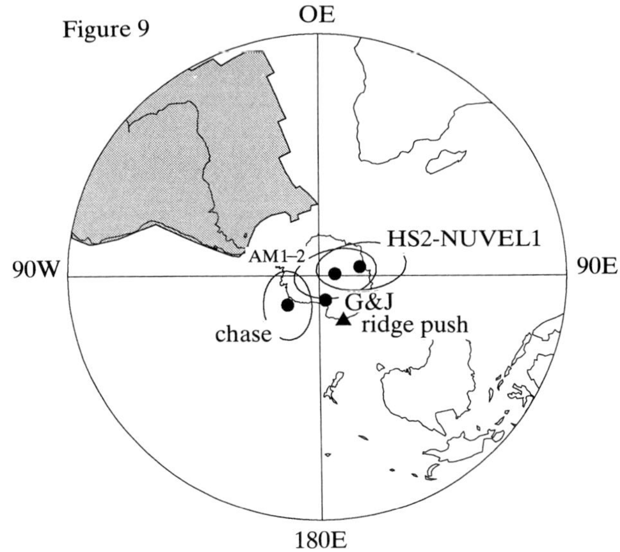

Figure 8. Stress magnitudes in the central Indian Ocean part of the Indo-Australian plate (grey shading). Plotted are the differential stresses at epicentral locations (see Govers *et al.* 1991) inferred from the hypocentral depths (see text). The ends of the thin solid lines at the base of the columns (if not present, the middle points of the base of the columns) indicate the epicentral locations. All stress values are expressed for a uniform reference value (100 km) of the plate thickness. The dark band in the column corresponds to the uncertainty in the inferred differential stress. The thicker horizontal line in the column gives the differential stress calculated by Cloetingh & Wortel (1985, 1986). (i) Stress from modelling ($\sigma_1 - \sigma_3$) from earthquakes.

Figure 9. Relation between driving forces and absolute motion for the South American plate. Circles: Poles of absolute-motion vectors for the South American plate, according to models AM1-2 (Minster & Jordan 1978), Gordon & Jurdy (1986, labelled G&J), Chase (1978), HS2-NUVEL1 (Gripp & Gordon 1990; see also DeMets *et al.* 1990). Triangle: pole of integrated torque vector of driving force (ridge push).

oceanic intraplate standards, and detailed studies of focal mechanisms are available (Wiens & Stein 1983, 1984; Bergman & Solomon 1984, 1985), and (2) for this plate Cloetingh & Wortel (1985, 1986) have calculated the stress field using the same procedure as outlined here.

For the Indo-Australian plate the agreement between calculated stress orientations and focal mechanism data is excellent (see Cloetingh & Wortel 1986; Bergman 1986). Also the orientations of observed long-wavelength basement undulations in the northeastern Indian Ocean appear to be consistent with calculated stress orientations (Stein *et al.* 1989).

(ii) *Magnitude*

Direct measurements of magnitudes of lithospheric stress are difficult to obtain. In a recent study we proposed a method to infer magnitude information from the depth of earthquake hypocentres. Using a lithospheric rheology model we relate the hypocentral depths of earthquakes quantitatively to the stress level in the lithosphere (see Govers *et al.* 1991). The depth-dependent rheology encompasses Byerlee's (1978) law, with the assumption of hydrostatic pore pressure, and flow laws derived for wet olivine (Tsenn & Carter 1987). Through the temperature-dependence of the flow laws, the dependence of the lithosphere's rheology on age is implicitly taken into account.

Since the strength value is taken to be the upper limit of the differential stress which can be supported elastically, Govers *et al.* (1991) adopted as a starting point of the analysis the fact that earthquakes do not occur at a certain depth if locally the stress is below the strength value. Hence, seismic activity indicates that the strength is reached. This enabled them to derive a minimum value for the differential stress ($\sigma_1 - \sigma_3$) from the depth of an earthquake's hypocentre in combination with the depth-dependent strength distribution. Uncertainties in focal depth, in age of the lithosphere and in strain rate (both of which affect the temperature-dependent ductile portion of the strength envelope), and in the slope of the brittle part of the strength envelope result in an uncertainty in the inferred value for the differential stress. The method was applied using the earthquakes of the Central Indian Ocean lithosphere for which accurate depth determinations are available (Wiens & Stein 1983, 1984; Bergman & Solomon 1984, 1985). Figure 8 shows the range of differential stress inferred from the seismicity data and also the differential stress calculated by Cloetingh & Wortel (1986).

With the exception of some events in the northeastern part of the Central Indian Ocean the agreement between seismicity derived values and model calculations is good. Stress levels about an order of magnitude lower, as advocated by Richardson (1989), are not sufficient to explain the observed seismicity, at least not according to the model (and the rheologies) used. We note, however, that relating earthquake occurrence to lithospheric stress requires a model or hypothesis for the relation between the two, which at present is not well established.

5. Discussion

(*a*) *Orientation of* $S_{\mathrm{H,max}}$ *and direction of absolute motion*

Zoback *et al.* (1989) draw attention to the observation that for some plates there is a tendency for the orientation of $S_{\mathrm{H,max}}$ (maximum compressive stress) to be (sub)parallel to the direction of absolute motion, and subsequently discussed possible mechanisms which may account for this feature. From our analysis we arrive at the following preferred interpretation of this observation (and implicitly also of deviations from this pattern). The driving forces govern the direction of absolute motion (see figure 6). The resistive forces passively adjust themselves to this imposed direction. The stress field, on the other hand, is a function of the entire set of forces acting on the plate. Depending on the geometry of the plate and the distribution of plate boundaries involving subduction zones and spreading ridges the principal-stress axes may align with the direction of the small circles of the driving forces. Thus, such an alignment is not a fundamental feature of the stress field. For the Pacific plate, for example, this alignment tendency is weak and it breaks down for the western part of the plate roughly coinciding with the lithosphere older than 80 Ma. We attribute this to the configuration of the convergent plate boundaries in the northwestern part of the Pacific plate which are located not only along its western boundaries but also along the northern edge of the plate.

(*b*) *Level of stress*

Upon comparison of figures 5 and 8 we note that the model results for the Indo-Australian plate gave a considerably higher level of intraplate stress than our results for the Pacific plate. Above we have presented evidence in support of the high level of stress in the Indo-Australian plate. We note that the Pacific plate results are of the same level of magnitude as earlier calculations of the stress level in the Nazca plate (Wortel & Cloetingh 1985; see also Richardson & Cox 1984). The difference in stress level between the Pacific plate, on the one hand, and the Indo-Australian plate on the other can be accounted for by the very different dynamic configuration of the plates involved. The Indo-Australian plate is characterized by high, and strongly non-uniformly distributed, resistance associated with continental collision, both in the Himalayan region and in the Banda arc. The Pacific plate does not encounter such high resistance, and the resistance is distributed more evenly. Hence, the slab pull forces can be compensated in a much smoother way thereby leading presumably to a situation with an approximate terminal velocity for subducting slabs, as invoked by Forsyth & Uyeda (1975). The same holds for the Nazca plate.

(*c*) *Other plates*

Through the choice of the Pacific plate the emphasis has been on oceanic lithosphere. The dynamics of this plate are strongly dominated by the slab-pull forces associated with the very long convergent plate boundaries. This renders the system insensitive to the magnitude of the smaller forces. The South American plate provides an interesting complementary situation. The ridge push is likely to be the main driving force, whereas the slab pull plays a very minor role (Lesser Antilles and South Sandwich subduction zones). Meijer & Wortel (1991) studied the dynamics of this plate in terms of a first-order model incorporating a ridge-push force $\boldsymbol{F}_{\mathrm{rp}}$, a drag force at the base of the lithosphere $\boldsymbol{F}_{\mathrm{dr}}$ (σ_{b}), a transform fault resistance $\boldsymbol{F}_{\mathrm{tf}}$, an upper plate resistance $\boldsymbol{F}_{\mathrm{up}}$ at the interface between the descending Nazca plate and the South American plate, and a resistive force associated with the contact between the South American plate and the Caribbean plate (called $\boldsymbol{F}_{\mathrm{Car}}$). For several segments of the plate's boundaries complications arise, the associated forces of which are largely unknown. They are dealt with in terms of superpositions onto the first-order model. Parameters specifying these additional forces (magnitude, azimuth) are determined from the torque balance constraint. Meijer & Wortel (1991) show that a large set of force distributions satisfies this constraint. This illustrates the need to consider specific aspects of an individual plate. By modelling the stress field (work in progress) it should be possible to discriminate between this large number of force distributions by using the observed stress directions.

A test similar to the one for the Pacific plate in figure 6, using Gordon *et al.*'s (1978) hypothesis, has been made for the South American plate. The magnitude of the ridge-push torque is 11.1×10^{25} N m, comparable with that of the Pacific plate (see table 2). In the absence of a significant slab-pull torque, however, the magnitude of the total torque of the driving forces is an order of magnitude smaller for the South American plate than for the Pacific plate. Nevertheless, the agreement between poles of absolute motion and the pole of the torque of the driving forces is good (see figure 9).

6. Conclusions

From our analysis of the Pacific plate we draw the following conclusions which we consider to be of general interest in the context of modelling dynamics and stress fields of the lithosphere.

1. Modelling the dynamics of the lithosphere via plate-tectonic forces satisfactorily

accounts for the direction of absolute motion, and provides a good basis for quantifying the contribution of tectonic forces to the intraplate stress field.

2. The slab pull acting on the subducted lithosphere is not necessarily compensated locally. When integrated over the full length of convergent plate boundaries, however, it is very nearly compensated by resistive forces. This lack of local compensation strongly affects the stress distribution in the plate.

3. Our numerical results show a preference for shear stresses at the base of the oceanic lithosphere to be resisting, rather than driving, plate motion.

4. A positive correlation between the orientation of $S_{\mathrm{H,\,max}}$ and the direction of absolute motion, which has been observed for some plates, is not a fundamental feature of the intraplate stress field. The entire set of forces acting on a plate should be considered. Depending on the nature and distribution of these forces an alignment of the two directions may or may not occur.

References

Artyushkov, E. V. 1973 The stresses in the lithosphere caused by crustal thickness inhomogeneities. *J. geophys. Res.* **78**, 7675–7708.

Bergman, E. A. 1986 Intraplate earthquakes and the state of stress in oceanic lithosphere. *Tectonophys.* **132**, 1–35.

Bergman, E. A. & Solomon, S. C. 1984 Source mechanisms of earthquakes near mid-ocean ridges from body waveform inversion: implications for early evolution of oceanic lithosphere. *J. geophy. Res.* **89**, 11415–11441.

Bergman, E. A. & Solomon, S. C. 1985 Earthquake source mechanisms from body-waveform inversion and intraplate tectonics in the northern Indian Ocean. *Phys. Earth planet. Int.* **40**, 1–23.

Byerlee, J. D. 1978 Friction of rocks. *Pageoph.* **116**, 615–626.

Chapple, W. M. & Tullis, T. E. 1977 Evaluation of the forces that drive the plates. *J. geophys. Res.* **82**, 1967–1984.

Chase, C. G. 1978 Plate kinematics: the Americas, East Africa, and the rest of the world. *Earth planet. Sci. Lett.* **37**, 355–368.

Cloetingh, S. & Wortel, R. 1985 Regional stress field of the Indian plate. *Geophys. Res. Lett.* **12**, 77–80.

Cloetingh, S. & Wortel, R. 1986 Stress in the Indo-Australian plate. *Tectonophys.* **132**, 49–67.

Crough, S. T. 1975 Thermal model of oceanic lithosphere. *Nature, Lond.* **256**, 388–390.

DeMets, C., Gordon, R. G., Argus, D. F. & Stein, S. 1990 Current plate motions. *Geophys. J. Int.* **101**, 425–478.

England, P. & Wortel, R. 1980 Some consequences of the subduction of young slabs. *Earth planet. Sci. Lett.* **47**, 403–415.

Fleitout, L. & Froidevaux, C. 1982 Tectonics and topography for a lithosphere containing density heterogeneities. *Tectonics* **1**, 21–56.

Forsyth, D. & Uyeda, S. 1975 On the relative importance of the driving forces of plate motion. *Geophys. Jl R. astr. Soc.* **43**, 163–200.

Forte, A. M. & Peltier, W. R. 1987 Plate tectonics and aspherical Earth structure. *J. geophys. Res.* **92**, 3645–3679.

Froidevaux, C., Uyeda, S. & Uyeshima, M. 1988 Island arc tectonics. *Tectonophys.* **148**, 1–9.

Goetze, C. & Evans, B. 1979 Stress and temperature in the bending lithosphere as constrained by experimental rock mechanics. *Geophys. Jl R. astr. Soc.* **59**, 463–478.

Gordon, R. G., Cox, A. C. & Harter, C. E. 1978 Absolute motion of an individual plate estimated from its ridge and trench boundaries. *Nature, Lond.* **274**, 752–755.

Gordon, R. G. & Jurdy, D. M. 1986 Cenozoic global plate motions. *J. geophys. Res.* **91**, 12389–12406.

Govers, R., Wortel, M. J. R., Cloetingh, S. A. P. L. & Stein, C. A. 1991 Stress magnitude estimates from earthquakes in oceanic plate interiors. *J. geophys. Res.* (In the press.)

Gripp, A. E. & Gordon, R. G. 1990 Current plate velocities relative to the hotspots incorporating the NUVEL-1 global plate motion model. *Geophys. Res. Lett.* **17**, 1109–1112.

Hager, B. H. & O'Connell, R. J. 1981 A simple model of plate dynamics and mantle convection. *J. geophys. Res.* **86**, 4843–4867.

Jacoby, W. R. 1970 Instability in the upper mantle and global plate movements. *J. geophys. Res.* **75**, 5671–5680.

Larson, R. L., Pitman III, W. C., Golovchenko, X., Cande, S. C., Dewey, J. F., Haxby, W. F. & LaBreque, J. C. 1985 The bedrock geology of the World. San Francisco: W. H. Freeman.

Lister, C. R. B. 1975 Gravitational drive on oceanic plates caused by thermal contraction. *Nature, Lond.* **257**, 663–665.

McKenzie, D. P. 1969 Speculations on the consequences and causes of plate motion. *Geophys. Jl R. astr. Soc.* **18**, 1–32.

McNutt, M. 1987 Lithospheric stress and deformation. *Rev. Geophys.* **25**, 1245–1253.

Meijer, P. Th. & Wortel, M. J. R. 1991 The dynamics of motion of the South American plate. *J. geophys. Res.* (In the press.)

Minster, J. B. & Jordan, T. H. 1978 Present-day plate motions. *J. geophys. Res.* **83**, 5331–5354.

Oxburgh, R. & E. M. Parmentier 1977 Compositional and density stratification in oceanic lithosphere. *J. geol. Soc. Lond.* **133**, 343–355.

Richardson, R. M., Solomon, S. C. & Sleep, N. H. 1979 Tectonic stress in the plates. *Rev. Geophys. Space Phys.* **17**, 981–1019.

Richardson, R. M. & Cox, B. L. 1984 Evolution of oceanic lithosphere: a driving force study of the Nazca plate. *J. geophys. Res.* **89**, 10043–10052.

Richardson, R. M. 1989 The origin of the intraplate stress field. *28th Int. Geol. Congress, Washington, Abstr. Vol.* **2**, 695–696.

Richter, F. M. & McKenzie, D. P. 1978 Simple plate models of mantle convection. *J. Geophys.* **44**, 441–471.

Scotese, C. R., Gahagan, L. M. & Larson, R. L. 1988 Plate tectonic reconstructions of the Cretaceous and Cenozoic ocean basins. *Tectonophys.* **155**, 27–48.

Solomon, S. C. & Sleep, N. H. 1974 Some simple physical models for absolute plate motions. *J. geophys. Res.* **79**, 2557–2567.

Stein, C. A., Cloetingh, S. & Wortel, R. 1989 SEASAT-derived gravity constraints on stress and deformation in the northeastern Indian Ocean. *Geophys. Res. Lett.* **16**, 823–826.

Stein, S., Cloetingh, S., Wiens, D. A. & Wortel, R. 1987 Why does near ridge extensional seismicity occur primarily in the Indian Ocean? *Earth planet. Sci. Lett.* **82**, 107–113.

Tsenn, M. C. & Carter, N. L. 1987 Upper limits of power law creep of rocks. *Tectonophys.* **136**, 1–26.

vanden Beukel, J. 1990 Thermal and mechanical modelling of convergent plate boundaries (Ph.D. thesis, University of Utrecht). *Geologica Ultraiectina* **62**. (126 pages.)

Vigny, C., Ricard, Y. & Froidevaux, C. 1991 The driving mechanism of plate tectonics. *Tectonophys.* **187**, 345–360.

Vlaar, N. J. & Wortel, M. J. R. 1976 Lithospheric aging, instability and subduction. *Tectonophys.* **32**, 331–351.

Weissel, J. K. & Anderson, R. N. 1978 Is there a Caroline plate? *Earth planet. Sci. Lett.* **41**, 143–158.

Wiens, D. A. & Stein, S. 1983 Age-dependence of oceanic intraplate seismicity and the implications for lithospheric evolution. *J. geophys. Res.* **88**, 6455–6468.

Wiens, D. A. & Stein, S. 1984 Intraplate seismicity and stress in young lithosphere. *J. geophys. Res.* **89**, 11442–11464.

Wortel, R. & Cloetingh, S. 1981 On the origin of the Cocos-Nazca spreading center. *Geology* **9**, 425–430.

Wortel, R. & Cloetingh, S. 1983 A mechanism for the fragmentation of oceanic plates. *AAPG Mem.* **34**, 793–801.

Wortel, M. J. R. & Cloetingh, S. A. P. L. 1985 Accretion and lateral variations in tectonic structure along the Peru-Chile Trench. *Tectonophys.* **112**, 443–462.

Wortel, M. J. R. & Vlaar, N. J. 1988 Subduction zone seismicity and the thermo-mechanical evolution of downgoing lithosphere. *Pageoph.* **128**, 625–659.

Zoback, M. D., *et al.* 1987 New evidence on the state of stress of the San Andreas fault system. *Science, Wash.* **238**, 1105–1111.

Zoback, M. L., *et al.* 1989 Global patterns of tectonic stress. *Nature, Lond.* **341**, 291–298.

Discussion

J. Cartwright (*Imperial College, London, U.K.*). Why do there appear to be no examples of compressional reactivation of extensional faults on a passive margin by the ridge-push force? Is this because the magnitude of the ridge-push force is too small even on old divergent margins?

M. J. R. Wortel. On the basis of the level of stress associated with the ridge-push force combined with the knowledge now available on the strength of the lithosphere (including lithosphere with pre-existing faults) we indeed consider this to be the principal cause. In the context of fault reactivation and seismic activity along passive margins we expect, however, that sources other than the ridge-push force, such as deglaciation and sediment loading, have to be envisaged as well.

M. H. P. Bott (*Durham University, U.K.*). The chief problem seems to be in assessing the slab-pull force. Does Dr Wortel's formula take into account the length of the slab?

M. J. R. Wortel. In calculating the slab-pull force the length of the slab was taken into account according to equation (3) which results from integrating the density difference (between the slab and the surrounding mantle) from the trench downdip to a distance S_{sz}. This parameter S_{sz} represents the downdip length of the Wadati–Benioff seismic zone associated with subducted oceanic lithosphere. A relation was derived from a global survey of subduction zones, and used in this paper, that S_{sz} varies with age t of the downgoing lithosphere according to $S_{sz} = 0.12\, v_c\, t$, where v_c is the convergence rate (with proper allowance for oblique convergence); see Wortel & Vlaar (1988) for a review. The above procedure implies that the slab-pull force in our modelling incorporates the anomalous density in the cold descending slab down to the depth of the deepest earthquake hypocentres in the subduction zone.

The forces driving the plates: constraints from kinematics and stress observations

By Donna M. Jurdy[1] and Michael Stefanick[2]

[1] *Department of Geological Sciences, Northwestern University, Evanston, Illinois 60208-2150, U.S.A.*

[2] *Science Systems and Applications, Inc., Seabrook, Maryland 20766, U.S.A.*

Plate kinematics and stress observations are used to assess the nature and relative magnitudes of the forces driving the plates. Dynamical equilibrium for a purely oceanic plate determines the relative magnitudes of the active and responding forces, largely balancing drag against slab-pull. Continental plates, moving more slowly, have errors of about 30% of the estimated motions. A torque balance model is used to describe the evolution of plate dynamics over the Cenozoic Era for reconstructions of plate geometry and velocities. Torques have been fairly stable for the past 64 Ma; the misfit to the model systematically increases for earlier times, most likely due to errors in the locations of past convergent plate boundaries and velocities. Unlike kinematics, which integrates the forces acting on a plate, the stress field responds locally and can differentiate between models for forces. Stress models which incorporate the forces are compared with stress orientations for the North and South American plates.

1. Introduction

What drives the plates? Do they drive themselves by subsidence and subduction of cooling oceanic lithosphere, or are they passively carried by mantle convection? The most direct evidence is kinematics, the plate motions themselves. We begin by looking at present kinematics realizing that, though this will give insight into dynamics, the insight fails in details. Starting with the present-day motions we define their characteristics, then discuss plate reconstructions over the Cenozoic Era. These snapshots of past plate geometries and velocities are important in testing hypotheses and models derived to explain present-day kinematics. In the next section we describe a general model to balance the forces on individual oceanic plates, and present an approach to convert to torques geometrically. The torque-balance model is applied to plate reconstructions using the reconstructed velocities at subduction zones to infer ages, and, consequently, the forces acting on individual plates. The torques are fairly stable and the global fit suggests that the model is generally valid. The difficulty with inferences about driving forces from kinematics is that the motion of a rigid plate represents an integrated effect of the torques acting on the plate. Different combinations of forces can give the same net motion. Furthermore, there is considerable variation among plate-motion models, particularly for the slower-moving continental plates. On the other hand, stress observations have the potential to give detailed information about the forces driving the plates, in that sets of forces with the same net torque can produce quite different stress patterns. For specified

Phil. Trans. R. Soc. Lond. A (1991) **337**, 127–139

Printed in Great Britain

force models the predicted stress fields can be determined and compared with observations of the World Stress Map (Zoback *et al.* 1989), even if the observations are only directional. Of course, stress magnitudes would further constrain models. In the final section we discuss stress observations and their utility as a test of dynamical models. The South American and North American plates are used as examples, because these plates are not dominated by long attached slabs.

2. Kinematics

Plate tectonics describes the relative motions of the plates on a global scale. Relative plate motions averaged over the past few million years, referred to as 'current' or 'present', are accurately known; locations and azimuths of transform faults and lineated magnetic anomalies, as well as slip vectors determined from analysis of earthquake focal mechanisms, give information about the relative motion of each of two adjacent plates. Data from all plate pairs can be combined to yield a self-consistent set of globally best-fitting angular-velocity vectors (Chase 1978*a*; Minster & Jordan 1978; DeMets *et al.* 1990). These relative motions are interpreted as 'absolute' motions with the assumption of a reference frame, such as hotspots assumed fixed relative to the mantle, or one in which there is 'no-net-rotation' of the set of motions. Investigations of present-day absolute plate motions have revealed certain patterns: the speed of a plate varies inversely with its continental area (every large oceanic plate moves faster than every large continental plate except for the Indian plate) (Minster *et al.* 1974; Forsyth & Uyeda 1975; Chapple & Tullis 1977), the speed of a plate increases with the fraction of its boundary that is being subducted (Forsyth & Uyeda 1975), speeds in equatorial regions tend to be greater than speeds in polar regions (Solomon *et al.* 1975), and overthrust plates at subduction zones all have a trenchward component of absolute motion, when behind-arc spreading is taken into account (Chase 1978*b*).

Reconstructions of plate velocities for earlier times are necessary to test these observations derived for the present. Global reconstructions of plate velocities in the past can be made using only seafloor-spreading data, and thus are not over-determined as is the present. These reconstructions are made in a circuit in which relative motions are determined for two plates separated by a ridge, and then one or both of these plates is similarly linked to other plates until all plates are related. For Cenozoic time global plate reconstructions (Gordon & Jurdy 1986) relate the Pacific-basin plates using seafloor spreading data and position them relative to the Pacific hotspots; similarly, the Atlantic plates are related and then fixed relative to the African hotspots. The assumption that the African and Pacific hotspots are fixed relative to each other requires motion along an arbitrary boundary in the Pacific plate. A reconstruction for the interval from 10 Ma to the present is shown in figure 1; the reconstruction for 64–56 Ma is shown in figure 2. This reconstruction and analysis of Cenozoic plate motions, relative to the hotspot and the mean-lithosphere reference frames, shows that several simple generalizations based on present plate motions have been valid over Cenozoic time. The tendency for plates to move faster near the equator than near the poles has persisted, with only slight variation, throughout the era. However, during the Early Cenozoic the Kula and Farallon plates – both at high latitudes – moved at speeds in excess of 120 mm a^{-1}. The inverse correlation between absolute velocity and continentality, characteristic of the present and the Palaeocene, holds in the intervening intervals. Generally plates

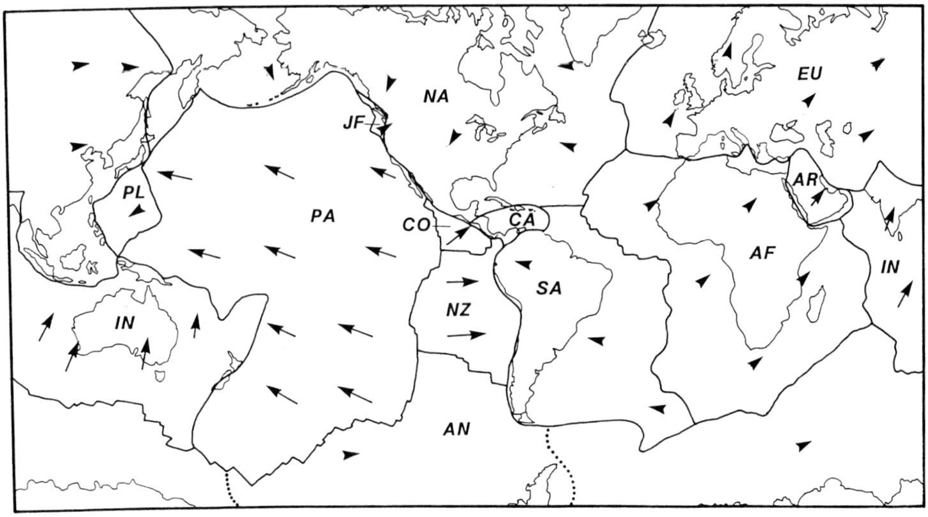

Figure 1. Cenozoic plate reconstruction, interval 10–0 Ma with present plate boundaries (from Gordon & Jurdy 1986). The arrow lengths at grid points are scaled for a constant time interval of 15 Ma. Mercator projection with top border at 70° N, bottom border at 70° S, and left and right borders at 90° E. Plate codes: AF, African; AN, Antarctic (East); AR, Arabia; AU, Australia; CA, Caribbean; CO, Cocos; CR, Chatham Rise (South Pacific); EU, Eurasian; FA, Farallon; IN, Indian; JF, Juan de Fuca; KU, Kula; LH, Lord Howe; NA, North American; NZ, Nazca; PA, (North) Pacific; PL, Philippine; PH, Phoenix (Aluk); and SA, South American.

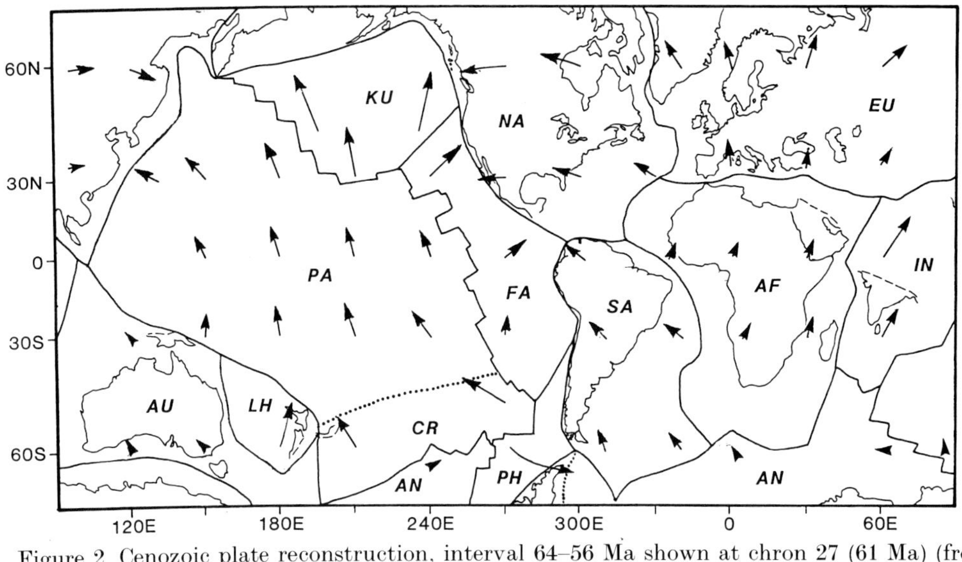

Figure 2. Cenozoic plate reconstruction, interval 64–56 Ma shown at chron 27 (61 Ma) (from Gordon & Jurdy 1986). The map parameters, arrow lengths and plate codes are as in figure 1.

moved rapidly toward their subduction zone. From this we conclude that the force of slab-pull is very important in driving the plates.

3. Dynamics

(a) *Dynamical equilibrium*

Plate motions themselves provide some information about the relative magnitudes of the forces and, as soon as motions were defined globally, dynamical models were attempted. Several authors found that pull on the downgoing slabs and total drag resistance dominate the balance (Harper 1975; Forsyth & Uyeda 1975; Chapple & Tullis 1977). Forsyth & Uyeda found that no other forces could be resolved from present plate motions; however, others (e.g. Elsasser 1971; Chapple & Tullis 1977; Chase 1978*a*) concluded that a net trenchward pull on overthrust plates may be significant at subduction zones. Also, Forsyth & Uyeda as well as Chapple & Tullis found that drag beneath the continents may be considerably stronger than drag beneath the oceans, though Harper's models (1975, 1986) show no such difference. Thus these early studies from kinematics come to a variety of conclusions about the relative importance of the forces driving the plates, but the studies are all based on a dynamical equilibrium between active and passive forces.

(b) *Two-dimensional model for oceanic plates*

A more recent simple dynamical model for oceanic plates (Carlson *et al.* 1983*a*, *b*; Carlson 1983) uses slab pull and ridge push as active forces to balance plate drag and slab resistance as response forces, ignoring other forces. Each force component is expressed as a function of age, speed, and length; a reasonable set of assumptions leads to a single equation with two parameters relating age and speed. From the relation for the age and speed, and then balancing plate drag against the slab pull and ridge push (Carlson *et al.* 1983*a*, *b*; Carlson 1983):

$$v^2t = Svt^{\frac{3}{2}} + Rt, \tag{1}$$

where v and t are the velocity and age, respectively, at the trench and S and R are coefficients fitted to 15 trenches around the world. Speeds and ages from the trenches are used to estimate the parameters, with the result that the ridge-push force is much smaller than the slab-pull force and may actually be zero within the errors (figure 3*a*). For a two-dimensional model of a plate there are three conditions for dynamical equilibrium; the net forces and the net torque about any point must all be zero. We used these conditions (Stefanick & Jurdy 1991) to estimate the magnitudes of forces as a function of age, with ridge push and slab pull as active forces, and plate drag and slab resistance as passive forces (figure 3*b*). The balance refers to purely oceanic plates where slab pull dominates. For continental plates, such as the case of South America to be discussed below, slab pull is less than ridge push.

(c) *Three-dimensional models for forces and torques*

The motion of a rigid plate over the Earth's surface has three degrees of freedom, rather than six degrees of freedom if it were free to move through space. Rigid-plate motion is completely described by its angular velocity vector $\boldsymbol{\omega}$. It seems reasonable to assume that we have a nearly perfect dynamical equilibrium with the passive torques (e.g. plate drag, slab resistance, collision) adjusting to balance the active torques (e.g. slab pull, ridge push, trench suction).

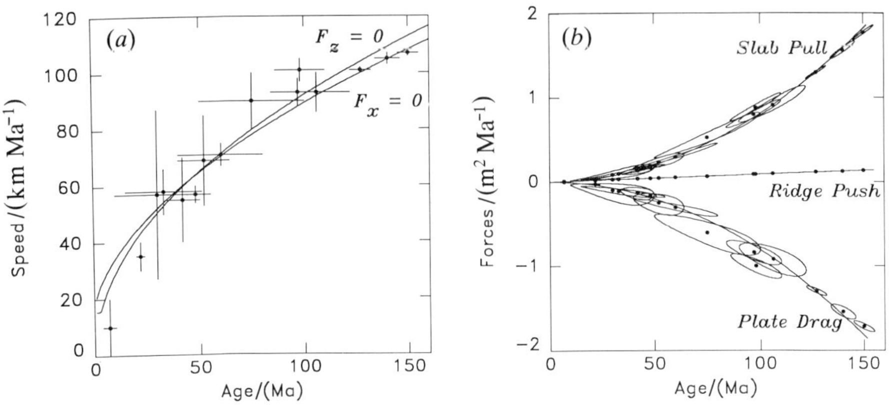

Figure 3. (*a*) Velocity–age relation for the plates at present and (*b*) force budget, with values shown for 15 trenches (from Stefanick & Jurdy 1991). (*a*) Median rate of motion against median age at 15 trenches and ranges of values (after Carlson *et al.* 1983*a*, fig. 2). The two curves correspond to the net vertical force equals zero and the net horizontal force equals zero. (*b*) Force balance against age showing slab pull, ridge push and plate drag, together with the error regions corresponding to the error bars in *a*. (Data points from Carlson *et al.* 1983*a*, fig. 3.)

The plate-drag torque $\boldsymbol{\tau}_D$ is assumed proportional to the angular momentum of the plate:

$$\boldsymbol{\tau}_D = -C_D I_p \boldsymbol{\omega}_p, \tag{2}$$

where I_p is the moment of inertia tensor of the plate, $\boldsymbol{\omega}_p$ is the plate's angular velocity, and C_D is the constant of proportionality. The drag model is the simplest one possible, ignoring any counterflow beneath the plate. The remaining torques are all along plate boundaries, and if the force is locally normal to the boundary and proportional to the element length,

$$\boldsymbol{\tau} = \int \boldsymbol{r} \times \mathrm{d}\boldsymbol{F} = C \int \boldsymbol{r} \times \left[\frac{-\boldsymbol{r} \times \mathrm{d}\boldsymbol{r}}{r_E}\right] = C r_E \int \mathrm{d}\boldsymbol{r} = C r_E \boldsymbol{r}_c, \tag{3}$$

where r_E is the radius of the Earth, $\boldsymbol{r}_c$ is the vector from the initial to the final point on the boundary segment (chord vector), and C is a constant of proportionality.

The slab-pull torque is proportional to the chord vector of the boundary and also depends on the age of the slab and the subduction rate (Carlson *et al.* 1983*a*, *b*; Harper 1975, 1986):

$$\boldsymbol{\tau}_{SP} = C_S |v_n| t^{\frac{3}{2}} \boldsymbol{r}_c, \tag{4}$$

where C_S is the constant of proportionality, v_n is the velocity component normal to the boundary, $\boldsymbol{r}_c$ is the chord vector, and t is the age at the trench at the midpoint of the chord. The ridge-push torque is simply proportional to the age:

$$\boldsymbol{\tau}_{RP} = C_R t \boldsymbol{r}_c. \tag{5}$$

At a convergent boundary it seems physically reasonable and necessary to include collision forces which will act outward from the boundary. These collision forces would assume whatever values are necessary to achieve a torque balance but must be equal in magnitude and opposite in direction, much as internal forces inside a plate adjust in such a way that the plate moves rigidly. How should we model these?

Table 1

NA	latitude	longitude	rate (deg Ma^{-1})	SA	latitude	longitude	rate (deg Ma^{-1})
MJ	−58.3	−40.7	0.247	MJ	−82.3	75.7	0.285
C	−36.8	−70.7	0.251	C	−70.7	−131.3	0.232
GG	−67.2	−11.1	0.280	GG	−70.3	74.7	0.320
DM	−39.8	−61.4	0.246	DM	−81.6	24.9	0.293

Using equation (1) we obtain the vector form:

$$I_{\mathrm{p}}\,\boldsymbol{\omega} = (S|v_{\mathrm{n}}|\,t^{\frac{3}{2}}/r_{\mathrm{E}}^{2})\,\Delta \boldsymbol{r} + R(t/r_{\mathrm{E}}^{2})\,\Delta \boldsymbol{r}. \qquad (6)$$

The simplest dynamical model for a continental plate without a large subduction zone is one that would balance plate drag against ridge push alone (and so in equation (6) the first term on the right would be dropped). In table 1 the absolute motions computed by using this model are compared with the estimates of Minster & Jordan (1978), Chase (1978*a*), Gripp & Gordon (1990), who computed absolute motions for NUVEL (DeMets *et al.* 1990), (abbreviated MJ, C, and GG, respectively) for North America and South America.

DM is a dynamical model that balances plate drag against ridge push alone and the rough agreement suggests that ridge push alone makes the major contribution to the active forces on these two plates. The link between relative and absolute plate motions is often based on hotspots and the assumption that they represent components of a single fixed set. However, hotspots may be deflected by the plate motions (Petronotis & Jurdy 1990) and in this respect dynamical model estimates may prove to be useful in defining an 'absolute' framework.

4. Cenozoic evolution of torques

For our previous study (Jurdy & Stefanick 1988) of driving forces over the Cenozoic, we used the reconstructions of Gordon & Jurdy (1986). In their analysis the Cenozoic Era is divided into six intervals; instead of regularly spaced time divisions, the times of major plate reorganizations were used as natural temporal boundaries: 64–56 Ma, 56–48 Ma, 48–43 Ma, 43–25 Ma, 25–10 Ma, and 10 Ma–present. The chron numbering convention and age assignments in millions of years of Harland *et al.* (1982) are followed. The absolute angular velocities of each plate relate to the hotspots for the six time intervals and the geometric factors characterizing the individual plates for each reconstruction were tabulated (Gordon & Jurdy 1986, tables 3 and 4). We used these in our computations of the torques acting on the plates (Jurdy & Stefanick 1988). Drag torque is found from the product of the moment of inertia tensor and the observed angular rotation vector. Balanced torque is found from an angular velocity vector which would exactly balance drag with active torques. These two torques are compared plate by plate and a global comparison is made by summing over the plates. (A full description is given by Jurdy & Stefanick (1988), eqs 12–15.) The global torque balance is shown for the interval 10 Ma to the present in figure 4 and for 64–56 Ma in figure 5. The misfit in torque, almost constant

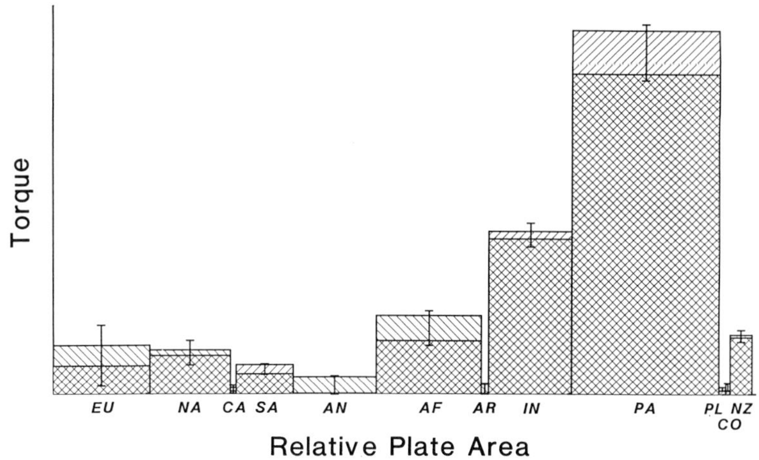

Figure 4. Torque comparison for individual plates between balanced torque and drag torque for the interval 10 Ma to present shown in figure 1. Plate velocities are in the hotspot framework (Gordon & Jurdy 1986, table 3). Torques obtained by solving the balancing equation (6) are shown by positively sloped lines, and the drag torque by negatively sloped lines. The misfit between the two torques is indicated as error bars about the mean. (From Jurdy & Stefanick 1988.)

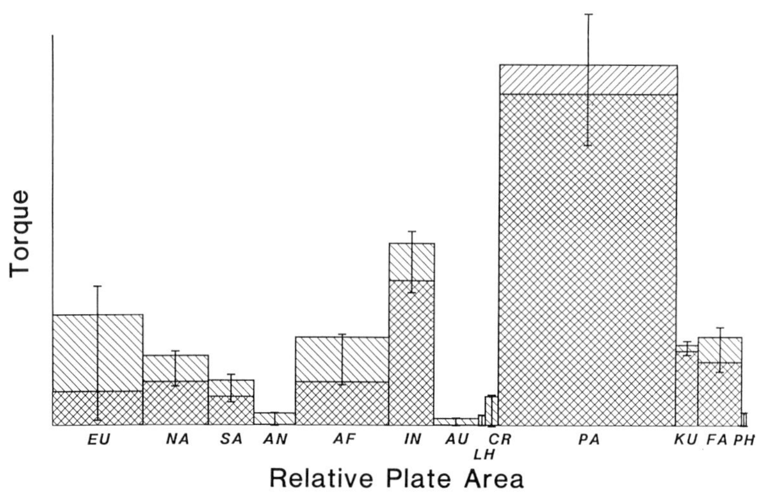

Figure 5. Torque comparison for individual plates between balanced torque and drag torque for the interval 64–56 Ma shown in figure 2. Plate velocities are in the hotspot framework (Gordon & Jurdy 1986, table 3). The torques and misfits are depicted as in figure 4. (From Jurdy & Stefanick 1988.)

for the past 43 Ma, shows an elbow and then a linear increase back in time to 64 Ma (figure 6). Alternatively, the misfit could be described as a hyperbola as shown by the dashed curve in figure 6. The hyperbola corresponds to the mean square error being the sum of two independent parts, one a constant (model misfit?), and the other proportional to $(\text{time})^2$, i.e. a systematic error growing with age (errors in reconstructions for boundaries and plate motions?). A generalization of this one-dimensional force model to two dimensions yields a torque model which is fairly stable over the Cenozoic Era.

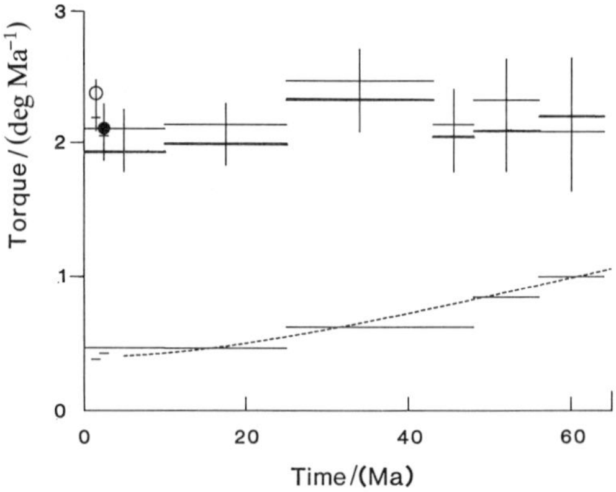

Figure 6. Drag torque and balance torque magnitudes for present and Cenozoic time, with the line segments corresponding to the six time intervals used for reconstructions (Gordon & Jurdy 1986). Drag torque for the present is shown as the open circle for the plate motion model of Minster & Jordan (1978) and the solid circle for the model of Chase (1978*a*); for display these are arbitrarily displaced from the ordinate. The upper sequence shows $|\tau_B|$, $|\tau_D|$, and $|\epsilon|$ with time; $|\epsilon|$ is shown with vertical bars about the mean torque value; the heavier of the lines is the drag torque. The lower sequence shows the evolution of the misfit $|\epsilon|$ as a solid line; the dashed curve is the hyperbola showing a systematic increase in the r.m.s. error with time. (From Jurdy & Stefanick 1988.)

5. Two-dimensional model with stress

The motion a rigid plate averages of the torques acting on the plate various combinations of forces can result in the same net motion. However, stress responds more locally, and sets of forces with the same net torques can produce stress patterns that are quite different. Recent compilations of stress measurements show fields changing in character over an individual plate and a tendency for the maximum horizontal stress direction, $S_{H,max}$ to be parallel to the absolute plate motion (Zoback *et al.* 1989). These observations of stress orientations and magnitudes allow detailed comparison with the predictions of models for various combinations of forces driving the plates. In our study we presented (Stefanick & Jurdy 1991) a set of models for possible forces driving the South American plate and compared them with observed stress measurements. The South American plate was chosen for study because the slab-pull component should be small and the other components are reasonably isolated geographically. Thus the plate balance is largely between ridge push and plate drag, with the opportunity of resolving the effects of trench suction or collision at the subduction zones. We estimated the ridge-push force to be about 5×10^{13} N m^{-1}, where the ocean floor is oldest, with an estimated age of 144 Ma, and the lithosphere has a thickness of 6×10^4 m. If the ridge-push force is balanced by drag acting over the entire plate, and the proportion of oceanic plate is 50%, then the stress due to the net force is a maximum at the midpoint and is about 500 MPa. Two small slabs supply extra forces, at the Scotia and Caribbean arcs, contributing about 20% of the ridge-push force. Most remarkable is the nearly uniform E–W orientation of the $S_{H,max}$ directions in the regional stress field (figure 7), which at high altitudes are extensional. The two-dimensional horizontal stress patterns are compared for successive models of combinations of driving forces. Models with ridge push, slab pull at the Scotia and Caribbean arcs, and trench suction near the west coast, all balanced by plate drag, produce similar stress patterns to those observed (Zoback *et al.* 1989). A model for the stress orientations for the North American plate

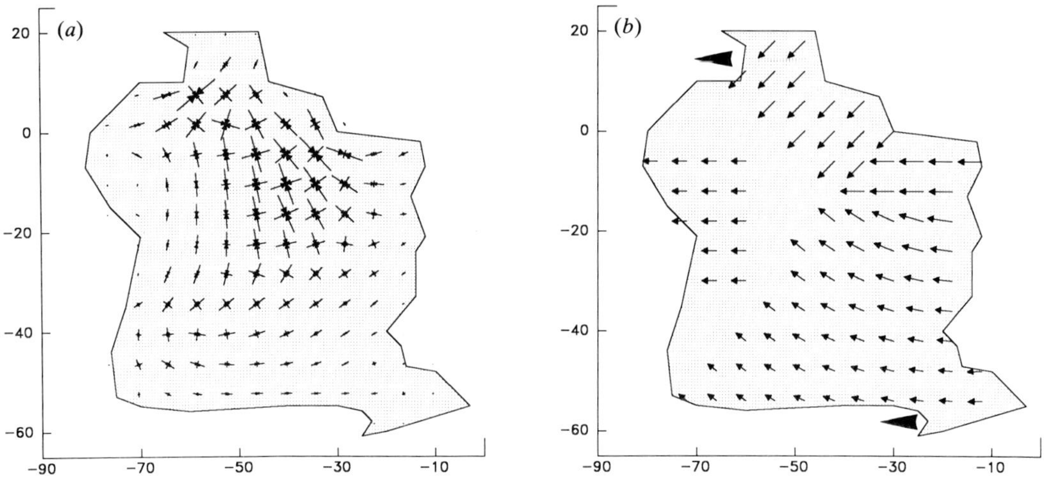

Figure 7. Stress pattern for the South American plate resulting from ridge push and slab pull (heavy arrows at the Scotia and Caribbean arcs), as well as trench suction at the western boundary, balanced by plate drag. Latitude and longitude are used as cartesian coordinates. (*a*) Principal horizontal stresses shown with outward arrows indicating tension and inward arrows for compression. Every other grid point shown. (*b*) Forces acting on plate shown as arrows at grid points.

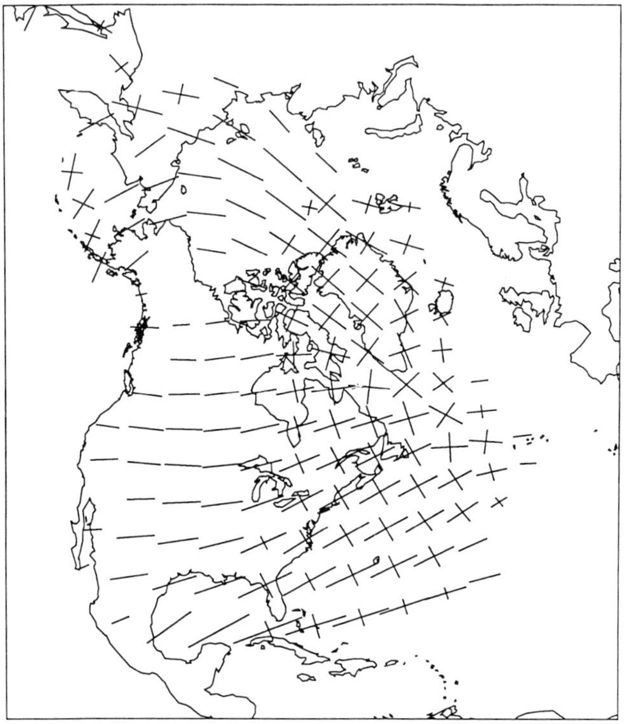

Figure 8. Stress orientation for the North American plate resulting from ridge-push and collision force at Aleutian arc. The line segments indicate directions of compression on a logarithmic scale. Extension not shown.

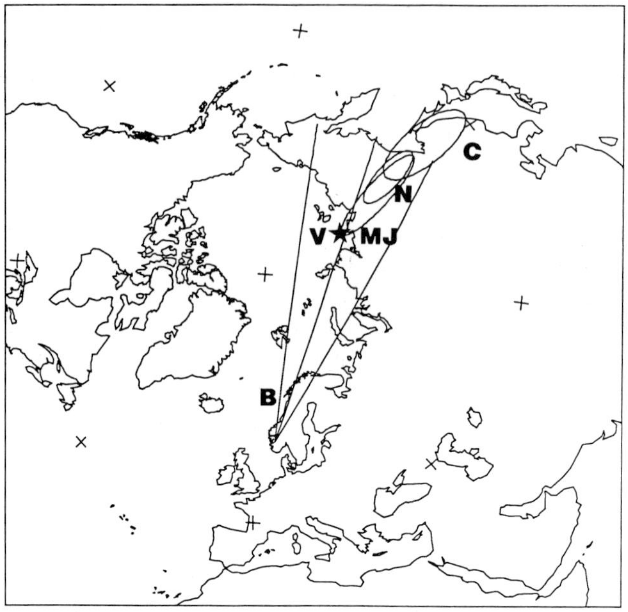

Figure 9. The pole position ellipses for the relative motion of NA and EU for the following plate motion models: MJ, Minster & Jordan (1978); C, Chase (1978); N, DeMets *et al.* (1990). The trend lines, B, are great circle arcs through Norway showing the range of stress directions from earthquakes (Bungum *et al.* 1991) and the pole positions as V for VLBI measurements of Herring *et al.* (1986).

is shown in figure 8, for the case of ridge push, with a small amount of collision force at the Aleutian arc.

The force field acting on a plate is a combination of the body and boundary force densities. Ridge push is a body force acting over the oceanic part of the plate, whereas plate drag is a passive body force acting over the entire plate. Slab pull is treated as a boundary force, as is collision. We find the force field for the combination of the forces hypothesized to be acting. Using this force field we calculate the stress tensor field and also the directions of the two principal horizontal stresses; these define the stress régime when compared with the vertical stress. Directions and stress régime are compared with observations.

For simplicity a planar model is described. The force (f_x, f_y) and torque M_z due to ridge push (and trench suction or collision) are calculated and the plate velocity (v_x, v_y) and angular velocity ω_z, required so that the net plate drag force and torque will balance, are calculated. The stress field is then calculated from the momentum equations (iteratively using a relaxation method on the displacements)

$$\partial\tau_{xx}/\partial x + \partial\tau_{xy}/\partial y + f_{ax} + f_{px} = 0, \tag{7}$$

$$\partial\tau_{yx}/\partial x + \partial\tau_{yy}/\partial y + f_{ay} + f_{py} = 0, \tag{8}$$

and boundary conditions. Here $\tau_{xx}, \tau_{xy}, \tau_{yx}, \tau_{yy}$ are the stress components, f_{ax}, f_{ay} the components of the active force per unit area and f_{px} the drag force per unit area. The boundary conditions specify the stress values at the appropriate points and these must be consistent with the net force calculations or the iteration misfit will not converge to zero. We treat the collision as a given (active) force rather than a passive (response) force. This is because we are treating the plate by itself. In a global model the collision force would adjust itself and so would be a passive force analogous to drag. Comparison of the derived kinematics and deformation with our dynamical models, including stress, should constrain possible force magnitudes and distributions on the plates. Observations from the World Stress Map will provide additional constraints, in that quite different stress patterns can result from sets of forces that produce the same net motion.

Norway has a low level of seismicity, with only a few earthquakes with magnitude greater than 5. Bungum *et al.* (1991) studied the focal mechanisms of Norway and argue that their solution indicates that the maximum compressive stress direction is due to ridge push. However, the maximum stress direction is only roughly parallel to the pole they use at 52° N, 126° E for the ridge-push force. Previous estimates of the rotation for Eurasia to North America at 62° N, 135° E are based on average motions over the past 3–5 Ma. In fact, the pole improved with VLBI measurements comes closest to fitting the current stress directions, as indicated by earthquake focal mechanisms shown in figure 9 by great-circle arcs through Norway.

6. Discussion

Present-day kinematic models show that oceanic plates move most rapidly and in the direction of the attached subducting slabs. Continental plates move more slowly and away from their ridges, but the errors in their motions are perhaps 30 % of the motion. Absolute plate motion models suggest that oceanic plates are driven by a combination of slab pull and ridge push, balancing plate drag. These force models must be first converted to torque models because plates are spherical shells moving over the Earth's surface. Since these forces are different functions of age and speed, coefficients are determined and these are used to estimate the relative magnitudes of these forces. The same coefficients can then be used to estimate the motion for continental plates and this works well for North and South America but not for Eurasia. Physically, we should expect a collisional force between Eurasia and the Indian and Pacific plates and this force is not modelled yet, but seems to balance the ridge push from the Atlantic. Modelling collision force is a prerequisite for more sophisticated models.

Plate reconstructions provide an important test for kinematic and dynamic models. Velocities at subduction zones are reconstructed and from these the ages and forces acting on the plates are inferred. Using the provisional force estimates, and converting these to torques, we find that global torque is nearly constant throughout the Cenozoic. For the earlier reconstructions the misfit to the model increases, probably due to uncertainties in locations of subduction zones and the velocities there. The global fit over the Cenozoic suggests that the force balance model is generally valid.

Stress observations, even if only directional, can be compared with the output of stress models using different combinations of boundary and body forces. At this stage conclusions can be only provisional, but for South America a combination of ridge push and slab pull (from two small arcs) balancing plate drag and some small collision along the west coast seems to give a more realistic model of the observed stress orientations than other combinations of forces. Stress responds locally to forces on the plate, so in detail, the stress orientations cannot always be in the direction of absolute plate motions. More accurate modelling of the ridge-push field and an allowance for lateral inhomogeneity, particularly for continental regions, would seem to be necessary as the next step. Stress magnitudes would further constrain driving force models, but these observations are at present very limited. However, some inferences can be made from the observed stress régimes about the relative magnitudes of the horizontal and vertical stresses: non-extensional focal mechanisms in the Andes at depths of more than 30 km correspond to a lower bound of 830 MPa for the maximum horizontal stress. Stress orientations, as estimated from earthquakes for the area near Norway, seem to be aligned along the small circles of the relative pole vector of North America and Eurasia which brings us full circle to the kinematics. In summary, stress observations in combination with stress models will improve our estimates of the relative importance of the forces (torques) driving the plates.

This research was supported by NASA project NAG5-885. We thank the organizers of the Discussion Meeting on tectonic stress in the lithosphere: Professor M. H. P. Bott, Dr J. D. Fairhead, Professor N. Kusznir and Dr R. B. Whitmarsh. We are grateful to Professor N. Kusznir for his review of our typescript.

References

Bungum, H., Kvamme, L. B. & Hansen, R. A. 1991 Seismicity and seismotectonics of Norway and nearby continental shelf. *J. geophys. Res.* **96**, 2249–2265.

Carlson, R. L. 1983 Plate motions, boundary forces, and horizontal temperature gradients: implications for the driving mechanism. *Tectonophys.* **99**, 149–164.

Carlson, R. L., Hilde, T. W. C. & Uyeda, S. 1983*a* The driving mechanism of plate tectonics: relation to age of the lithosphere at trenches. *Geophys. Res. Lett.* **10**, 297–300.

Carlson, R. L., Hilde, T. W. C. & Uyeda, S. 1983*b* Correction to 'The driving mechanisms of plate tectonics: Relation to the age of lithospheric trenches'. *Geophys. Res. Lett.* **10**, 987.

Chapple, W. M. & Tullis, T. E. 1977 Evaluation of the forces that drive the plates. *J. geophys. Res.* **82**, 1967–1984.

Chase, C. G. 1978*a* Plate kinematics: the Americas, East Africa, and the rest of the world. *Earth planet. Sci. Lett.* **37**, 355–368.

Chase, C. G. 1978*b* Extension behind island arcs and motions relative to hot spots. *J. geophys. Res.* **83**, 5385–5387.

DeMets, C., Gordon, R. G., Argus, D. F. & Stein, S. 1990 Current plate motions. *Geophys. J. Int.* **101**, 425–478.

Elsasser, W. M. 1971 Seafloor spreading as thermal convection. *J. geophys. Res.* **76**, 1101–1112.

Forsyth, D. W. & Uyeda, S. 1975 On the relative importance of the driving forces of plate motion. *Geophys. Jl R. astr. Soc.* **43**, 163–200.

Gordon, R. G. & Jurdy, D. M. 1986 Cenozoic global plate motions. *J. geophys. Res.* **91**, 12389–12406.

Gripp, A. E. & Gordon, R. G. 1990 Current plate velocities relative to the hotspots incorporating the NUVEL-1 global plate motion model. *Geophys. Res. Lett.* **17**, 1109–1112.

Harland, W. B., Cox, A. V., Llewellyn, P. G., Pickton, C. A. G., Smith, A. G. & Walters, R. 1982 *A geologic time scale.* (131 pages.) Cambridge University Press.

Harper, J. F. 1975 On the driving forces of plate tectonics. *Geophys. Jl R. astr. Soc.* **40**, 465–474.

Harper, J. F. 1986 Mantle flow and plate motions. *Geophys. Jl R. astr. Soc.* **87**, 155–171.

Herring, T. A. *et al.* 1986 Geodesy by radio interferometry: evidence for contemporary plate motion. *J. geophys. Res.* **91**, 8341–8347.

Jurdy, D. M. & Stefanick, M. 1988 Plate driving forces over the Cenozoic Era. *J. geophys. Res.* **93**, 11833–11844.

Minster, J. B. & Jordan, T. H. 1978 Present-day plate motions. *J. geophys. Res.* **83**, 5331–5354.

Minster, J. B., Jordan, T. H., Molnar, P. & Haines, E. 1974 Numerical modelling of instantaneous plate tectonics. *Geophys. Jl R. astr. Soc.* **36**, 541–576.

Petronotis, K. E. & Jurdy, D. M. 1990 Pacific plate reconstructions and uncertainties. *Tectonophys.* **182**, 383–391.

Solomon, S. C., Sleep, N. H. & Richardson, R. M. 1975 On the forces driving plate tectonics: inferences from absolute plate velocities and intraplate stress. *Geophys. Jl R. astr. Soc.* **42**, 769–801.

Stefanick, M. & Jurdy, D. M. 1991 Stress observations and driving force models for the South American plate. *J. geophys. Res.* (In the press.)

Zoback, M. L. *et al.* 1989 Global patterns of tectonic stress. *Nature, Lond.* **341**, 291–298.

Discussion

M. H. P. Bott (*Durham University, U.K.*). Does Professor Jurdy have any handle on the viscosity underneath the plates, as it varies from one plate to another?

D. M. Jurdy. No, by using a linear drag model we are assuming a constant viscosity model. I do not think we want to add a lot of individual parameters at this stage.

M. J. R. Wortel (*Utrecht University, The Netherlands*). Is the misfit between the observations and the model predictions due to the relation used between the absolute velocity and the age of the lithosphere at the Trench?

D. M. Jurdy. Most of the misfit is due to uncertainties in the reconstruction, both for subduction velocities and plate geometry. These uncertainties are the portion that grow as we go back in time.

L. Fleitout (*École Normale Supérieure, Paris, France*). It can be argued that slab pull is proportional to the square root of lithospheric age and not to (age)$^{\frac{3}{2}}$. Would the uncertainty in the relationship used to calculate the slab-pull force have an important effect on these models and the calculated stress distribution?

D. M. Jurdy. The $vt^{\frac{3}{2}}$ law that is used for slabpull is generally agreed to be the correct one, and it gives good correlations for the present.

J. Cartwright (*Imperial College, London, U.K.*). Is there a correlation between the plates with large areas of continental lithosphere, which seem to move slower than the oceanic plates and the absence of a low velocity zone beneath cratonic cores?

D. M. Jurdy. At least for the Cenozoic Era, large continental plates lack subducting slabs and thus move more slowly than oceanic plates. This may not always have been true.

State of stress and crustal deformation along weak transform faults

By Mark D. Zoback

Department of Geophysics, Stanford University, Stanford, California 94305, U.S.A.

The state of stress and style of crustal deformation along weak transform plate boundaries is discussed in the context of available data and simple analytical models. Appreciable evidence indicates that while the frictional strength of the upper crust is high (in general accordance with simple faulting theory and laboratory-derived coefficients of friction), the frictional resistance to motion along transform plate boundaries is extremely low. These conditions require that horizontal principal stresses must be oriented approximately parallel and perpendicular to the transform-fault zone to minimize the shear stresses acting parallel to the transform. Along plate margins that must accommodate relative convergence (transpressive margins), a pattern of near-fault-normal compression and fault-normal crustal shortening is expected. Along divergent plate margins (transtensional margins), extension is expected to occur perpendicular to the transform, as the direction of minimum horizontal principal stress is expected to be nearly perpendicular to it. These patterns of stress and deformation can be observed along a number of transform faults around the world.

1. Introduction

In situ stress measurements to depths of about 3 km in reverse, strike–slip and normal-fault environments indicate states of stress consistent with Mohr–Coulomb frictional faulting theory and laboratory-derived coefficients of friction (Raleigh *et al.* 1972; McGarr & Gay 1978; Brace & Kohlstedt 1980; Zoback & Hickman 1982; Pine *et al.* 1983; Zoback & Healy 1984; Stock *et al.* 1985; Baumgärtner & Zoback 1989; Baumgärtner *et al.* 1990; Zoback & Healy 1991). These data would seem to argue that it is relatively simple to predict the state of stress in the lithosphere utilizing 'Byerlee's law' (Byerlee 1978) for the brittle upper crust and uppermost mantle and a power-law creep rheology (which is dependent on mineralogy, temperature and strain rate) for the ductily deforming lower crust and deeper mantle (Kirby 1980; Sibson 1982, 1983; Chen & Molnar 1983; Smith & Bruhn 1984; Molnar 1988).

However, in marked contradiction to Byerlee's law, a growing set of data indicates that motion along major plate-bounding transform faults like the San Andreas is resisted by extremely small shear stresses. These data include conductive heat-flow measurements in numerous, relatively shallow boreholes drilled along the length of the San Andreas fault but which detect no evidence of frictionally generated heat (see Brune *et al.* 1969; Henyey & Wasserburg 1971; Lachenbruch & Sass 1973, 1980, 1988). These data imply that the average frictional resistance to fault motion is no greater than about 20 MPa (Lachenbruch & Sass 1981). For the case of hydrostatic pore pressure, Byerlee's law predicts levels of shear stress of about 150 MPa at about 15 km for strike–slip faulting. In addition, data on the orientation of principal

Phil. Trans. R. Soc. Lond. A (1991) **337**, 141–150
Printed in Great Britain

horizontal stresses obtained from earthquake focal-plane mechanisms and borehole measurements in central California show that there is extremely small shear stress acting on planes parallel to the San Andreas fault because the direction of the maximum horizontal principal stress is nearly perpendicular to the strike of the fault (Zoback *et al.* 1987; Mount & Suppe 1987; Oppenheimer *et al.* 1988; Wong 1990). In a 3.5 km deep borehole drilled adjacent to the fault at Cajon Pass, California (a site presumably quite late in the earthquake cycle) a complete lack of right-lateral shear stress on planes parallel to the San Andreas fault was found (Shamir & Zoback 1991; Zoback & Healy 1991) even though the measured differential stress levels were consistent with Byerlee's law. These data indicate that while the frictional strength of faulted crust adjacent to the San Andreas fault is high, the frictional strength of the fault itself is quite low. Also in the Cajon Pass borehole, no evidence was found of fault-generated frictional heat, nor of heat redistribution due to thermal convection (Lachenbruch & Sass 1991), thereby substantiating the validity of the heat-flow measurements in shallow boreholes along the length of the fault. Finally, the near-orthogonal relationship between oceanic transform faults and oceanic spreading centres implies that the resistance to spreading is much greater than that to shear motion along the transforms (Lachenbruch & Thompson 1972; Oldenburg & Brune 1972, 1975). Wilcock *et al.* (1990) have recently shown that normal-faulting microearthquakes adjacent to the Kane transform fault have B and T axes that are approximately parallel and perpendicular to the transform.

2. A conceptual model of stress and deformation along weak transform faults

Taken together, the data referred to above suggest that faulting and crustal deformation along weak transform plate boundaries are controlled by a 'strong crust and weak transform'. Shear stress in the crust appears to be generally high (i.e. consistent with Byerlee's law), whereas shear stresses resolved on planes parallel to the transform faults appear to be quite low. Zoback *et al.* (1987) argued that, if the San Andreas fault was quite weak, principal stresses must be oriented parallel and perpendicular to it so as to result in low shear stress parallel to the transform. The simple cartoon in figure 1 represents the conceptual relationship between stress directions and deformation along a transform plate boundary that is considerably weaker than the adjacent crust. This cartoon is presented in the context of a map based on an oblique Mercator projection about the pole of relative plate motion. With such a projection, a transform fault that exactly coincides with small circles on a globe appear as a straight line parallel to the direction of relative motion. The purpose of figure 1 is to illustrate that for the case of a weak transform fault along a transpressive margin, one would expect a pattern of near-fault-normal compression and fault-normal crustal shortening. Along a weak divergent transform margin extension is expected to occur perpendicular to the transform and the direction of minimum horizontal principal stress is expected to be nearly perpendicular to it.

Figure 2 presents a map of maximum horizontal stress orientations along the San Andreas fault, Cerro Prieto in northern Mexico and the transform faults that form the boundary between the Pacific and North America plates in the Gulf of California. The map is plotted in an oblique Mercator projection using the pole of relative motion between the Pacific and North America plates after DeMets *et al.* (1990).

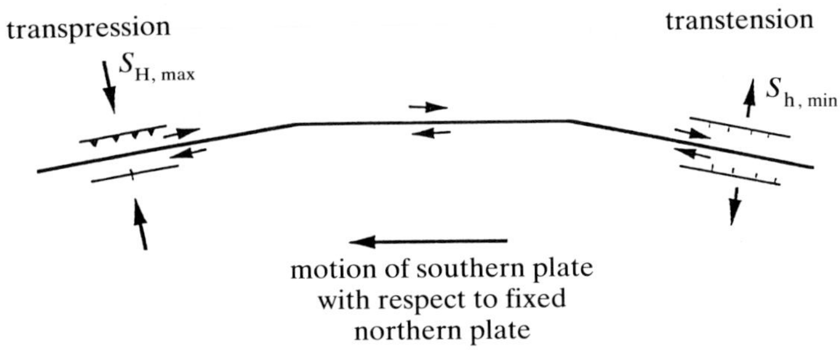

Figure 1. Cartoon illustrating the relationship between the state of stress and deformation along a transform fault that is much weaker than the surrounding crust. For comparison with figure 2, the cartoon is shown in the context of an oblique Mercator projection for the pole of relative motion.

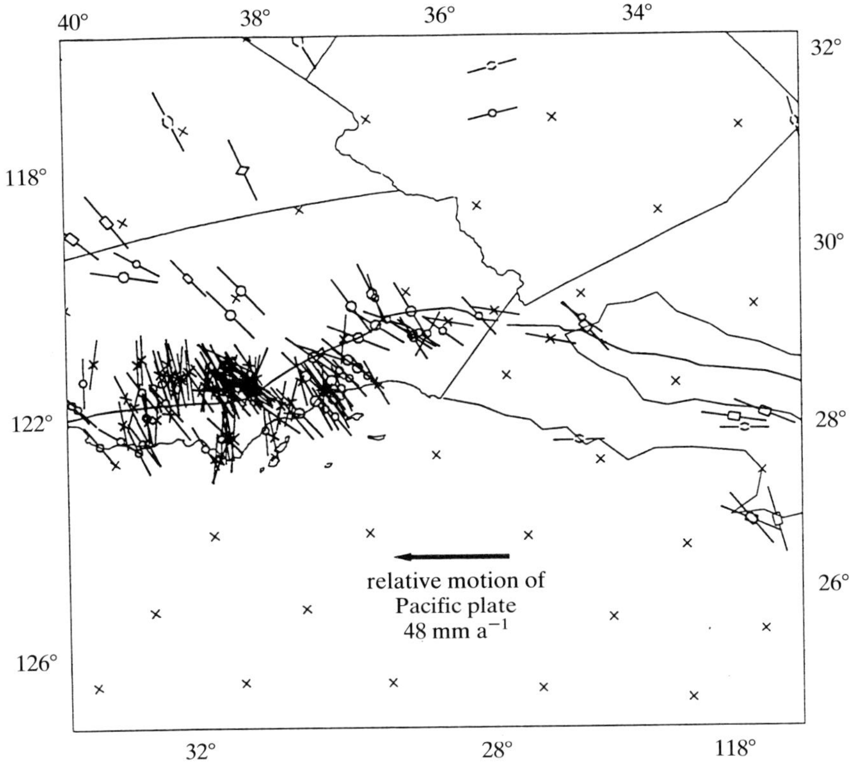

Figure 2. Stress map showing directions of maximum horizontal compression along the San Andreas, Cerro Prieto and Gulf of California transform faults. Map projected about the pole of relative motion between the Pacific and North America plates as given by DeMets *et al.* (1990). The symbol at the centre of each stress orientation indicates the type of stress indicator (circle, earthquake focal plane mechanism; arrow heads, wellbore breakouts; diamond, *in situ* stress measurement; square, geological indicator).

Because the data density is so high in central California only A and B quality data are shown on this map (see Zoback & Zoback 1989).

The manner in which the overall pattern of stress orientation and crustal deformation changes with the trend of the transform is remarkably consistent with the predictions of the simple cartoon shown in figure 1. Near-fault-normal compression is observed to the north of the eastern Transverse ranges in southern California where the plate motion is relatively convergent, and numerous active folds and reverse faults in the Coast Ranges that trend sub-parallel to the San Andreas fault (Page 1981) clearly indicate fault-normal crustal shortening as indicated in figure 1. Near fault-normal extension (which is shown in figure 2 as near-fault-parallel compression) is observed along southernmost San Andreas fault (near the Salton Sea and Imperial Valley), and along the Cerro Prieto fault and the transform faults in the Gulf of California. Similarly, along the Dead Sea transform in Israel the direction of minimum principal stress is sub-perpendicular to the strike of the fault and near-fault-normal extension is occurring (Eyal & Reches 1983; Letouzey 1986).

Note that transpression or transtension along a weak transform margin need not be constant through time as relative plate motion can change. Although relative plate motion along most of the San Andreas fault is currently convergent, from the late Miocene to Pliocene, tectonism adjacent to the fault was dominated by basin formation and transform-normal extension. About 4 Ma ago, the modern compressional deformation (the onset of folding and reverse faulting indicating shortening perpendicular to the San Andreas fault) coincided with a change in absolute motion of the Pacific Plate (Page & Engebretsen 1984). This change in plate motion resulted in relative-motion vectors changing from slightly divergent to slightly convergent. Similarly, along the Dead Sea transform Eyal & Reches (1983) and Letouzey (1986) discuss a transition from fault-normal compression observed before 8 Ma to the fault-normal extension observed today in the context of a change in relative plate motion.

3. Simple models of the state of stress along weak transform faults

Zoback and others (1987) presented a simple analysis of a weak strike–slip fault based on the assumption that the San Andreas fault zone represents a 'weak fault in a strong crust' and computed the orientation of the maximum principal stress near the fault as a function of the far-field stress orientation and average fault strength. They assumed that the magnitude of the far-field shear stress at depth could be computed from Byerlee's law and that the mean stress magnitude in the crust did not change as a function of distance from the fault. Figure 3 presents an extension of their analysis. Note that, except for the case where the far-field stress orientation is exactly 45° to the fault, a marked change of maximum principal stress direction occurs so as to minimize shear stress parallel to the weak transform. The sense of stress rotation (resulting in either fault-normal compression or fault-normal extension) depends on whether the far-field stress orientation corresponds to either convergent or divergent plate motion. The difference between the near-field and far-field angles depends only weakly on the average strength of the fault as long as it is considerably less than that of the crust far from the fault. As mentioned above an upper bound of 20 MPa for the average frictional strength of the fault is based on the absence of any detectable frictionally generated heat.

Although the calculations shown in figure 3 are based on an extremely simple model, they provide a conceptual framework for interpreting changes from convergent to divergent motion along the strike of a major transform (as illustrated in figure 1) and changes from transpressive to transtensional deformation through time in response to changes of relative plate motion such as observed above for both the San Andreas and Dead Sea transforms.

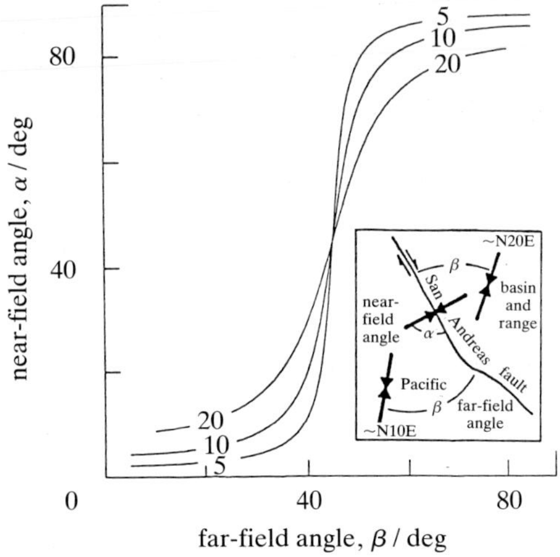

Figure 3. Orientation of maximum horizontal compression near a weak transform fault as a function of the far-field angle for different values of average fault strength (shown in megapascals).

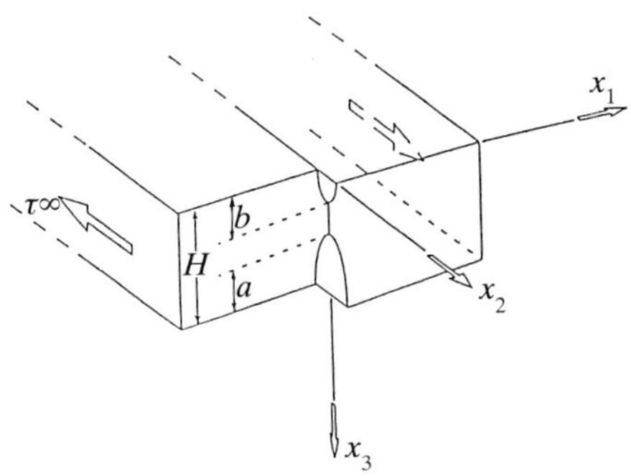

Figure 4. Model of strike–slip plate boundary after Tse *et al.* (1985) in which fault motion is driven by far-field shear τ^{∞} and occurs without frictional resistance over the depth ranges indicated by parameters a and b.

To improve upon the simple model of faulting along a weak transform described above, we consider now the state of stress in the vicinity of an extremely weak transform fault by adapting a model derived by Tse *et al.* (1985) that is illustrated in figure 4. In this model motion on a vertical transform fault is driven with a far-field shear stress, τ^{∞}. The angle between the strike of the fault and the far-field maximum horizontal principal stress is defined by the angle β. There is no frictional resistance to motion over depths defined by the parameters a or b, as defined in the figure. In the calculations below we have chosen $H = 100$ km to correspond to the lithosphere and we assume that there is no frictional resistance to sliding along the 'weak' transform throughout the entire crust and upper mantle. We thus set $b =$ 50 km and assume that $a = 0$ because we assume that at depths greater than 50 km both the fault and adjacent lithosphere have the same low strength (see below).

For the parameters defined above, figure 5*a* shows the calculated orientation of $S_{\mathrm{H,max}}$ at different depths in the upper crust for the case when relative plate motion is convergent (i.e. $\beta = 55°$). Fault normal compression is observed directly at the

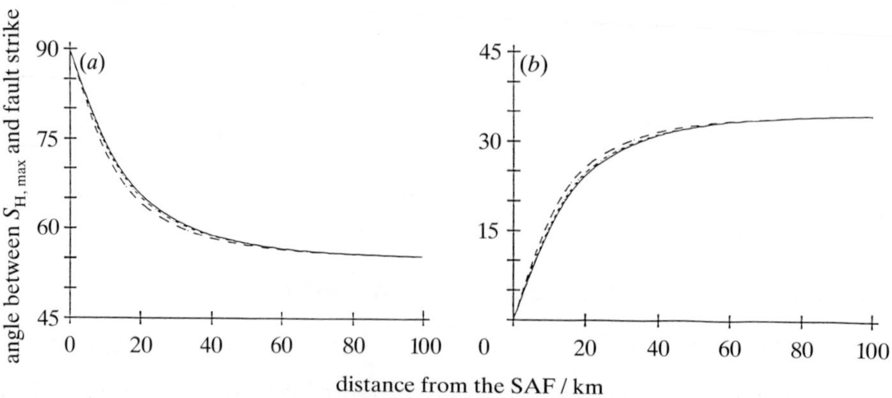

Figure 5. (*a*) Direction of maximum horizontal stress as a function of distance from the San Andreas fault (SAF) for different depths in the crust for the case of a convergent plate margin ($\beta = 55°$, $b = 50$ km). (*b*) Direction of maximum horizontal stress as a function of distance from the San Andreas fault for different depths in the crust for the case of a divergent plate margin ($\beta = 35°$, $b = 50$ km). (*a*, *b*) ——, $z = 3$ km; —·—·—, $z = 9$ km; — — —, $z = 15$ km.

fault (because it has zero frictional resistance) and the $S_{H, max}$ direction gradually approaches the far-field value over a distance comparable to b. The calculated $S_{H, max}$ direction is uniform throughout the upper crust. The case of divergent plate motion is shown in figure 5*b* (for $\beta = 35°$) where it is observed that the direction of maximum horizontal stress is subparallel to the fault at small distances and approaches the far-field stress orientation at distances comparable to b.

It is clear that although the simple model illustrated in figure 4 predicts fault-normal compression (or extension) near the fault, the change in stress orientation with distance is much more continuous and gradual than the broad (*ca.* 100 km) zone of near-fault-normal compression observed adjacent to the San Andreas fault (see Zoback *et al.* 1987). In a strict sense, the only way to explain this, in the context of the model shown in figure 4, would be to make the depth to which the weak fault extended much greater. A reasonable alternative to this hypothesis, however, is simply to consider the San Andreas fault system to be composed of a number of subparallel weak transform faults. Seismicity data clearly define such a series of faults throughout the Coast Ranges (see Hill *et al.* 1990).

4. Crustal decoupling

As originally pointed out by Lachenbruch & Sass (1973), equilibrium conditions require that the lack of frictional resistance along a vertical strike–slip fault be compensated by resistive basal stresses acting on subhorizontal planes adjacent to the fault. For the same model used for the calculations in figure 5*a*, the magnitudes of these drag forces are shown as a function of distance from the fault for different crustal depths in figure 6*a*. As indicated by the left ordinate of figure 6*a*, the calculations are presented as the basal shear normalized by τ^{∞}, the average far-field shear stress. It is clear that potentially large, resistive basal stresses are concentrated in the lower crust within distances to the fault comparable with b.

To compute the absolute magnitude of these resistive basal stresses it is necessary to estimate τ^{∞}. We thus require a far-field constitutive model for the crust and upper mantle along a strike–slip plate margin. We follow Kirby (1980), Sibson (1982, 1983) and others (see §1) and assume that the state of stress is controlled by Byerlee's law

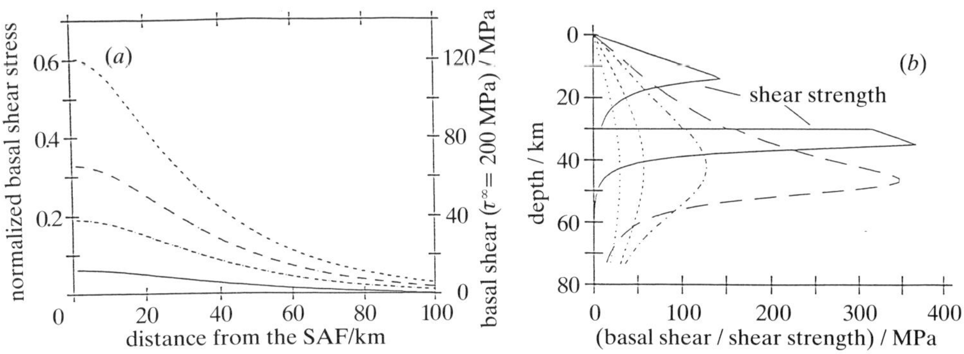

Figure 6. (*a*) Magnitude of resistive basal shear as a function of distance from the San Andreas fault for different depths in the crust. Left ordinate indicates magnitude of basal shear normalized by far-field driving shear stress, τ^∞. Right ordinate indicates magnitude of basal shear for τ^∞ = 200 MPa. ($\beta = 55°$, $b = 50$ km.) ——, $z = 3$ km; —·—·—, $z = 9$ km; ———, $z = 15$ km; ----, $z = 25$ km. (*b*) Magnitude of basal shear (for $\tau^\infty = 200$ MPa) as a function of depth at various distances from the fault and shear strength as a function of depth from figure 7 for a strain rate of 10^{-11} s^{-1} (see text). ——, $y = 5$ km; —·—·—, $y = 20$ km; ----, $y = 40$ km; ····, $y = 60$ km.

in the upper crust and uppermost mantle and by power-law creep in the lower crust and deeper mantle (figure 7). To arrive at the thickness-averaged far-field driving stress, τ^∞, we will simply divide the cumulative force required to cause deformation (the integral of the strength envelope) by the total thickness. Although it is quite difficult to attempt to calculate precise estimates of strength, several assumptions can be made which effectively result in a lower-bound strength estimate (choosing not to disregard brittle strength completely by assuming, for example, near-lithostatic pore pressure at depth). By multiplying τ^∞ by the normalized basal shear shown in figure 6*a*, we can then predict a minimum value for the basal shear resistance. The calculations of shear strength shown in figure 7 are done for a representative regional heat flow for western California (Lachenbruch & Sass 1973, 1980), a crustal thickness of 30 km and a mean normal stress acting on the transform fault that is approximately equal to the lithostatic stress (i.e. the far-field vertical stress is assumed to be approximately equal to the mean of the two horizontal principal stresses). Stresses are calculated for strain rates of 10^{-11}, 10^{-12} and 10^{-13} s^{-1}, which encompasses reasonable values for the plate-boundary region. An olivine composition is assumed for the mantle. For a heat flow of 71 mW m^{-2} the mantle has extremely low strength below 50 km, regardless of the strain rate. Two assumptions that are quite important with respect to computing a lower-bound estimate for τ^∞ are 1) a granitic composition for the lower crust (after Sibson 1982, 1983), which promotes creep at relatively shallow depth and 2) that hydrostatic pore pressure exists and that the effective-stress principal governs faulting in the brittle régimes throughout the upper crust and uppermost mantle.

For the assumptions stated above and far-field strain rates between 10^{-13} and 10^{-11} s^{-1}, the cumulative force needed to deform the crust and upper mantle is between about 5×10^{12} and 10^{13} N m^{-1}, corresponding to τ^∞ of 100–200 MPa averaged over a 100 km thick lithosphere. Multiplying this value by the normalized basal shear stress shown by the left ordinate of figure 6*a* yields the estimates of basal shear indicated by the right ordinate of figure 6*a* (the values shown correspond to τ^∞ of 200 MPa). If we had assumed that the composition of the lower crust was more

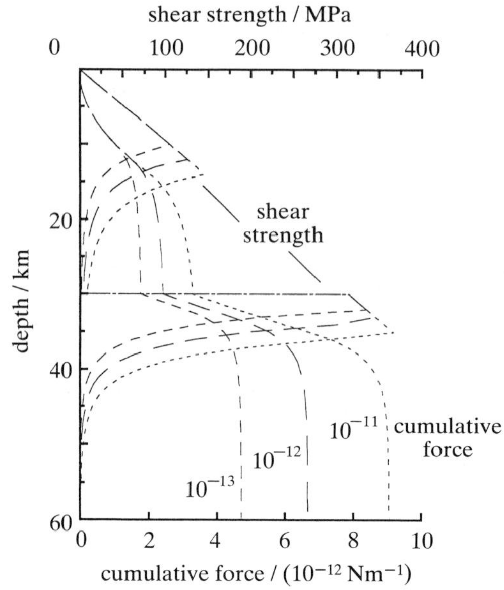

Figure 7. Shear strength and cumulative force as a function of depth and strain rate (s^{-1}) for a strike–slip faulting régime and a heat-flow characteristic of western California. Other parameters and assumptions are explained in the text.

mafic, or if the upper mantle had deformed under effectively 'dry' conditions (either because no pore fluid was present or because the effective stress principal did not apply) a greater cumulative force would have resulted which would cause the basal shear stresses shown in figure 6*a* to be even higher.

An important question about the magnitude of the basal shear stress is whether it is capable of causing faulting along horizontal planes. Because it was assumed that the normal stress on vertical strike–slip faults was approximately equal to the lithostat (figure 7), the strength envelopes used in that figure are also appropriate for a subhorizontal fault plane. In figure 6*b* we overlay the magnitudes of the basal shear at various depths with the strength envelope, for a strain rate of 10^{-11} s^{-1}, that was shown in figure 7. It can be seen that close to the fault the magnitude of basal shear exceeds the strength at depths greater than 15–20 km in the lower crust, and at depths beneath the strong mantle-lid. Thus, as predicted by Lachenbruch & Sass (1973), the motion of the upper crust along weak transforms like the San Andreas fault seems to be resisted by a decoupled lower crust.

References

Baumgärtner, J. & Zoback, M. D. 1989 Interpretation of hydraulic fracturing pressure–time using interactive analysis methods. *Int. J. Rock Mech. Min. Sci.* **26**, 461–470.

Baumgärtner, J., Rummel, F. & Zoback, M. D. 1990 Hydraulic fracturing in situ stress measurements to 3 km depth in the KTB pilot hole VB. A summary of a preliminary data evaluation. In *KTB Rep.* 90–6*a*, 353–400.

Brace, W. F. & Kohlstedt, D. L. 1980 Limits on lithospheric stress imposed by laboratory experiments. *J. geophys. Res.* **85**, 6248–6252.

Brune, J. N., Henyey, T. L. & Roy, R. F. 1969 Heat flow, stress, and rate of slip along the San Andreas fault, California. *J. geophys. Res.* **74**, 3821–3827.

Byerlee, J. D. 1978 Friction of rocks. *Pure appl. Geophys.* **116**, 615–629.

Chen, W. P. & Molnar, P. 1983 Focal depths of intracontinental and intraplate earthquakes and their implications for the thermal and mechanical properties of the lithosphere. *J. geophys. Res.* **88**, 4183–4214.

DeMets, D., Gordon, R. G., Argus, D. F. & Stein, S. 1990 Current plate motions. *Geophys. J. Int.* **101**, 425–478.

Eyal, Y. & Reches, Z. 1983 Tectonic analysis of the Dead Sea rift region since the Late-Cretaceous based on megostructures. *Tectonics* **2**, 167–185.

Henyey, T. L. & Wasserburg, G. J. 1971 Heat flow near major strike–slip faults in California. *J. geophys. Res.* **76**, 7924–7946.

Hill, D., Eaton, J. P. & Jones, L. M. 1990 Seismicity, 1980–1986. In *The San Andreas fault system, California* (ed. R. Wallace), pp. 115–152. U.S. Geological Survey Professional Paper 1515.

Kirby, S. 1980 Tectonic stresses in the lithosphere: constraints provided by the experimental deformation of rocks. *J. geophys. Res.* **85**, 6353–6363.

Lachenbruch, A. H. & Sass, J. H. 1973 Thermo-mechanical aspects of the San Andreas fault system. In *Proc. Conf. Tectonic Problems of the San Andreas Fault System* (ed. R. L. Kovach & A. Nur), pp. 192–205. Palo Alto: Stanford University Press.

Lachenbruch, A. H. & Sass, J. H. 1980 Heat flow and energetics of the San Andreas fault zone. *J. geophys. Res.* **85**, 6185–6223.

Lachenbruch, A. H. & Sass, J. H. 1981 Corrections to 'Heat flow and energetics of the San Andreas fault zone' and some additional comments on the relation between fault friction and observed heat flow. *J. geophys. Res.* **86**, 7171–7172.

Lachenbruch, A. H. & Sass, J. H. 1988 The stress heat-flow paradox and preliminary thermal results from Cajon Pass. *Geophys. Res. Lett.* **15**, 981–984.

Lachenbruch, A. H. & Sass, J. H. 1991 Heat flow from Cajon Pass, fault strength and tectonic implications. *J. geophys. Res.* (In the press.)

Lachenbruch, A. H. & Thompson, G. A. 1972 Oceanic ridges and transform faults: their intersection angles and resistance to place motion. *Earth planet. Sci. Lett.* **15**, 116–122.

Letouzey, J. 1986 Cenozoic paleo-stress pattern in the Alpine foreland and structural interpretation in a platform basin. *Tectonophys.* **132**, 215–231.

McGarr, A. & Gay, N. C. 1978 State of stress in the Earth's crust. *A. Rev. Earth planet. Sci.* **6**, 405–436.

Molnar, P. 1988 Continental tectonics in the aftermath of plate tectonics. *Nature, Lond.* **335**, 131–137.

Mount, V. S. & Suppe, J. 1987 State of stress near the San Andreas fault: implications for wrench tectonics. *Geology* **15**, 1143–1146.

Oldenburg, D. W. & Brune, J. 1972 Ridge transform fault spreading pattern in freezing wax. *Science, Wash.* **178**, 301–304.

Oldenburg, D. W. & Brune, J. N. 1975 An explanation for the orthogonality of ocean ridges and transform faults. *J. geophys. Res.* **80**, 2575–2585.

Oppenheimer, D. H., Reasenberg, P. A. & Simpson, R. W. 1988 Fault-plane solutions for the 1984 Morgan Hill, California earthquake sequence: evidence for the state of stress on the Calaveras fault. *J. geophys. Res.* **93**, 9007–9026.

Page, B. 1981 The southern Coast Ranges. In *The geotectonic development of California* (ed. W. G. Ernst). Englewood Cliffs, New Jersey: Prentice-Hall.

Page, B. & Engebretson, D. C. 1984 Correlation between the geologic record and computed plate motions in central California. *Tectonics* **3**, 133–155.

Pine, R. J., Ledingham, P. & Merrifield, C. M. 1983 *In situ* stress measurement in the Carnmenellis granite. II. Hydrofracture tests at Rosemanowes Quarry to depths of 2000 m. *Int. J. Rock Mech. Min. Sci.* **20**, 63–72.

Raleigh, C. B., Healy, J. H. & Bredehoeft, J. D. 1972 Faulting and crustal stress at Rangely, Colorado. In *Flow and fracture of rocks*, pp. 275–284. Geophysical Monograph, vol. 16. American Geophysical Union.

Shamir, G. & Zoback, M. D. 1991 A crustal stress orientation profile to 3.5 km depth near the San Andreas fault at Cajon Pass, California. *J. geophys. Res.* (Submitted.)

Sibson, R. H. 1982 Fault zone models, heat flow, and the depth distribution of earthquakes in the continental crust of the United States. *Bull. seismol. Soc. Am.* **72**, 151–163.

Sibson, R. H. 1983 Continental fault structure and the shallow earthquake source. *J. Geol. Soc. Lond.* **140**, 741–767.

Smith, R. B. & Bruhn, R. L. 1984 Intraplate extensional tectonics of the eastern Basin-Range: inferences on structural style from seismic reflection data, regional tectonics, and thermal-mechanical models of brittle–ductile transition. *J. geophys. Res.* **89**, 5733–5762.

Stock, J. M., Healy, J. H., Hickman, S. H. & Zoback, M. D. 1985 Hydraulic fracturing stress measurements at Yucca Mountain, Nevada, and the relationship to the regional stress field. *J. geophys. Res.* **90**, 8691–8706.

Tse, S. T., Dmoska, R. & Rice, J. R. 1985 Stressing of locked patches along a creeping fault. *Bull. seismol. Soc. Am.* **75**, 709–736.

Wilcock, W. S. D., Purdy, G. M. & Solomon, S. C. 1990 Microearthquake evidence for extension across the Kane transform fault. *J. geophys. Res.* **95**, 15439–15462.

Wong, I. 1990 Seismotectonics of the Coast Ranges in the vicinity of Lake Berryessa, northern California. *Bull. seismol. Soc. Am.* **80**, 935–950.

Zoback, M. D. & Healy, J. H. 1984 Friction, faulting and *in situ* stress. *Annls Geophisicae* **2**, 689–698.

Zoback, M. D. & Healy, J. H. 1991 *In situ* stress measurements to 3.5 km depth in the Cajon Pass Scientific Research Borehole: implications for the mechanics of crustal faulting. *J. geophys. Res.* (In the press.)

Zoback, M. D. & Hickman, S. 1982 *In situ* study of the physical mechanisms controlling induced seismicity at Monticello Reservoir, South Carolina. *J. geophys. Res.* **87**, 6959–6974.

Zoback, M. D., Zoback, M. L., Mount, V. S., Suppe, J., Eaton, J. P., Healy, J. H., Oppenheimer, D., Reasenberg, P., Jones, L., Raleigh, C. B., Wong, I. G., Scotti, O. & Wentworth, C. 1987 New evidence on the state of stress of the San Andreas fault system. *Science, Wash.* **238**, 1105–1111.

Zoback, M. L. & Zoback, M. D. 1989 Tectonic stress field of the conterminous United States. *Geol. Soc. Am. Mem.* **172**, 523–539.

Inferences of deviatoric stress in actively deforming belts from simple physical models

By Philip England[1] and Peter Molnar[2]

[1]*Department of Earth Sciences, Oxford University, Parks Road, Oxford OX1 3PR, U.K.*

[2]*Department of Earth, Atmospheric, and Planetary Sciences, Massachusetts Institute of Technology, Cambridge, Massachusetts 02139, U.S.A.*

Observations of temperatures and heat flux near major thrust zones indicate that in their deep levels shear stresses may exceed 50–100 MPa. Within strike–slip zones shear stresses in the lower lithosphere may also approach 50–100 MPa, though shear stresses in the upper crust of those regions are probably much lower. The relationship of tectonic style to surface elevation in the Andes and Tibet yields an estimate of about 5×10^{12} N m^{-1} for the force per unit length required to deform the lithosphere of these regions. This force per unit length is equivalent to an average shear stress of about 25 MPa through a lithosphere 100 km thick. The width-to-length ratios of active belts are consistent with deformation determined by the creep of the lower lithosphere rather than by friction on faults. The patterns of rotation of crustal blocks in western North America suggest that these blocks passively follow the deformation of a continuous substrate. The observations of deformation and the estimates of stress derived from them, both suggest that the upper continental crust is weak, relative to the lower parts of the lithosphere the deformation of which it follows passively. If this is so, determinations of stress in the upper crust may have only limited relevance to the deformation of the lithosphere as a whole.

1. Introduction

In contrast to the oceanic portions of the Earth's surface, which move as essentially rigid bodies, the continents exhibit distributed deformation over horizontal scales of hundreds to thousands of kilometres (Molnar & Tapponnier 1975; Molnar 1988*a*; England & Jackson 1989). A key problem in continental dynamics, therefore, is to relate the velocity gradients inferred from geodetic, seismological, palaeomagnetic, and structural data to the stresses acting within the lithosphere. The term 'lithosphere', as used here, refers to the crust and that part of the upper mantle through which heat is transferred primarily by conduction.

Orientations of principal stresses, and sometimes their magnitude can, in principle, be constrained by direct observation in boreholes in the uppermost crust (Brereton & Müller 1991; Fuchs *et al.* 1991; Zoback & Magee 1991; Zoback 1991). Such observations are usually expensive to make and, with a very few exceptions, are limited to depths of a few kilometres or less. The observation that strains and velocity gradients in the crust are organized on a horizontal lengthscale that far exceeds the thickness of the crust has led several authors to suggest that the upper crust passively follows the deformation of the lower lithosphere (Ekström & England 1989; England & Jackson 1989; England & McKenzie 1982; McKenzie & Jackson

Phil. Trans. R. Soc. Lond. A (1991) **337**, 151–164

Printed in Great Britain

1983; Jackson & McKenzie 1988). If this suggestion is correct, then the stress in the upper crust is not directly related to the state of stress in the deeper lithosphere (Houseman & England 1986, Appendix A).

The moment tensors of earthquakes give indications of strain (Kostrov 1974) through ranges of depths that are inaccessible to routine boring. These measurements of strain may, with much care, be used to supplement determinations of stress. But, because most of our information about active strain in the continents has come from the measurement of strain using the moment tensors of earthquakes (Ekström & England 1989; Jackson & McKenzie 1984, 1988; McKenzie 1978; Molnar & Deng 1984; Molnar & Lyon-Caen 1989; Molnar & Tapponnier 1975), any information on stress that might come from earthquake studies could not be used in a dynamical study without circularity.

Laboratory studies of the strain of rocks have been a major source of information on deformation processes in the inaccessible parts of the crust and mantle. The problems of extrapolating the results of these studies to geological conditions are too well known to need rehearsing here. Even in the absence of uncertainty in material properties determined in the laboratory, uncertainty in temperature and lithology at any given level within the lithosphere is sufficiently large to make a prediction of deviatoric stress from laboratory flow laws uncertain by an order of magnitude at least (see, for example, Molnar 1991).

The purpose of this paper is to review estimates of the magnitude of deviatoric stress in the deeper portions of the lithosphere that do not depend upon downwards extrapolation of observations made near the surface, nor upon using laboratory determinations of rheology. In the absence of a direct observation of stress in the deep lithosphere, these estimates must come from comparing the results of theory with observations of other quantities. The observations involved are heat flux and temperature near major shear zones, and the distributions of strain rate and topography in zones of active continental deformation. The theories, befitting our nascent understanding of the problem, are simple; they contain, in the first case, a single governing parameter and, in the second case, two.

2. Estimates of stresses from temperatures in major fault zones

Most of the energy dissipated by friction on faults is likely to be released as heat (see discussions by Lachenbruch & Sass (1980)). In aseismic shear zones, the rate of release of heat by dissipation is (neglecting the expenditure of energy in creating new surfaces) equal to the deviatoric stress multiplied by the strain rate. Observations of the thermal state of large faults and shear zones offer, therefore, the possibility of determining the magnitude of deviatoric stress within them.

(a) *Thrust faults*

Molnar & England (1990) gave simple formulae that permit estimates to be made of the deviatoric stress on thrust faults, either from surface heat flux measurements made above active thrust faults, or from temperatures inferred from exhumed thrust faults. The surface heat flux, Q, above a thrust fault that is slipping at a speed V is related to the shear stress, τ, on the fault and the heat flux, Q_0, that would be observed in steady state, without slip on the fault, by:

$$Q = (Q_0 + \tau V)/S \tag{1}$$

Table 1. *Estimates of shear stress in active subduction zones*

(z_f is the depth of the subduction zone below the site of heat flux measurement. V is the rate of subduction and τ is the shear stress calculated from equation (1) on the assumption that the shear stress is independent of depth; calculations assuming that shear stress increases proportionately to depth yield stresses approximately 30% higher (Kao & Chen 1991). The range in shear stress reflects uncertainties in the quantities entering equation (1).)

	z_f km	$V/(\text{mm a}^{-1})$	τ/MPa
Japan[a]	70 ± 10	105 ± 10	65–100
	60 ± 10	105 ± 10	55–90
S. Ryukyu–Kyushu[b]	25 ± 3	52 ± 13	35–50
N. Ryukyu–Kyushu[b]	10 ± 3	34 ± 12	25–50
	19 ± 5	34 ± 12	35–75
	30 ± 8	34 ± 12	80–115
Peru[a]	30 ± 5	80 ± 5	30–80
Himalaya[a]	15–18	15 ± 5	70–140

[a] Molnar & England (1990).
[b] Kao & Chen (1991).

(Molnar & England 1990, eq. (28)) where

$$S \approx 1 + \surd(Vz_f \sin\delta/\kappa) \tag{2}$$

and δ is the dip of the fault, z_f is the depth of the fault at the point of interest and κ is the thermal diffusivity of the rocks.

Molnar & England (1990) and Kao & Chen (1991) applied this analysis to estimate shear stresses in active subduction zones. The estimates lie in the range 25–140 MPa, and generally increase with the depth of the fault (table 1). Molnar & England (1990) used related arguments to estimate the shear stress on the active thrust fault that bounds the south of the Himalayas and obtained an estimate of 70–140 MPa for the shear stress at 15–18 km on that fault.

It appears from these estimates that shear stress in the upper portions of major active thrust faults is several tens to *ca.* 100 MPa. These stresses, resolved horizontally, contribute to the support of mountain ranges (Molnar & Lyon-Caen 1988). The range of shear stresses listed in table 1 for subduction zones implies a compressional force per unit length of convergent boundary of about $2.5 \times 10^{12}\ \text{N m}^{-1}$. The equivalent quantity calculated for the seismic part of the active intracontinental thrust fault beneath the Himalayas is 0.5 to $1 \times 10^{12}\ \text{N m}^{-1}$.

Estimates of driving forces based on shear stresses determined from the seismically active portions of faults are lower limits to the driving forces associated with major thrust faults, for they contain no contribution from stress on the deeper, aseismic, portions of those shear zones. Several authors have attempted to estimate shear stresses from temperatures inferred from metamorphic mineral assemblages generated during the slip of large thrust faults. In the studies summarized below, with the exception of that of Barton & England (1979), these mineral assemblages formed at temperatures above 450 °C and therefore above the assumed maximum temperature at which moderately sized-to-large earthquakes occur within continental crust (Chen & Molnar 1983). Analysis of such exhumed thrust faults can, therefore, provide information complementary to that derived from active fault zones.

The observation of an inverted metamorphic zonation associated with major thrust faults led to the suggestion that dissipative heating in the fault zones might be an important source of heat from metamorphism (Le Fort 1975). Early estimates of shear stress, from geological observations made on such faults, were around 100 MPa (Barton & England 1979; Graham & England 1976; Scholz 1980). These estimates were all based upon the steepness of the preserved metamorphic gradient near the fault, which was assumed to give an indication of the thermal gradient, and hence the rate of dissipative heating, near the fault (Graham & England 1976). The recognition that many zones showing inverted metamorphic zonation have been tectonically thinned after the metamorphism was imprinted (see, for example, England *et al.* 1991) removes the quantitative basis for those estimates. It is none the less clear from qualitative considerations that dissipative heating plays an important role in the thermal evolution of at least some thrust fault zones. In the absence of dissipative heating, the temperature at a particular point on the hanging wall of a thrust fault must drop once slip on the fault begins, because for most geologically relevant rates of slip, the temperature on a thrust fault that slips without dissipating appreciable heat is no higher than the average of the pre-slip temperatures of the rocks juxtaposed by slip on the fault. Because, in the cases discussed below, slip on the faults brought rocks originally at or close to the Earth's surface to the point of interest on the fault, and because the syn-slip temperatures recorded in the metamorphic assemblages approach or exceed the likely pre-slip temperatures in the hanging walls, some heat source – presumably dissipative heating – must have been operating at the faults.

Re-examination of the earlier estimates of shear stress from the Pelona Schist (Graham & England 1976) and from the Olympos Thrust (Barton & England 1979) using Molnar & England's (1990) approach, and taking account of post-metamorphic thinning of the inverted metamorphic sequence, yields estimates of shear stress that are somewhat lower than those obtained in the earlier studies, but still in the range 100–200 MPa (England & Molnar 1990).

From an analysis of the temperatures and pressures recorded by mineral assemblages generated during slip on the Main Central Thrust in the Annapurna–Manaslu region of central Nepal, England *et al.* (1991) concluded that shear stresses in the range 50–100 MPa were required to generate the inferred syntectonic temperatures on the Main Central Thrust. A similar magnitude of shear stress is obtained from Hubbard's (1989) inverted metamorphic sequence beneath the Main Central Thrust in the Mt Everest region (England & Molnar 1990).

If we combine the estimates of shear stress from the deeper, now inactive, Main Central Thrust and the estimate, made above on the basis of present day seismicity, of shear stress on the active thrust fault bounding the Himalayas, we may conclude that shear stress as high as 100 MPa acted to a depth of 35 km or more on the Main Central Thrust whilst it was active and, by extension, that a similar stress distribution is associated with the present thrust fault beneath the Himalayas.

Furthermore it seems reasonable to conclude, from estimates of shear stress at other major thrust faults, that the Main Central Thrust is not unusual in having supported shear stress of about 100 MPa during its slip. Shear stresses of this magnitude, resolved horizontally, would provide a driving force per unit length of about 4×10^{12} N m^{-1}, if the shear stress were independent of depth, or about 2×10^{12} N m^{-1} if it rose linearly from zero at the Earth's surface to about 100 MPa at a depth of 40 km.

(*b*) *Strike–slip zones*

The absence of a measurable heat flux anomaly associated with the San Andreas fault has been taken as evidence for generally low shear stresses on the fault (Brune *et al.* 1969; Lachenbruch & Sass 1980). In contrast, Scholz *et al.* (1979) concluded, from the distribution of metamorphism associated with the Alpine fault of the South Island of New Zealand – another major transcurrent zone – that this fault must have experienced shear stresses in excess of 100 MPa during a substantial portion of its slip.

Lachenbruch & Sass (1980) summarized over 90 heat flux determinations near the San Andreas fault and showed that there is no measurable heat flux anomaly associated with the fault itself and that, therefore, the shear stress on the fault is probably less than 10 MPa. On the other hand heat flux measurements in central California show an anomaly approximately 80 km wide, which covers the southern Coastal Ranges. Figure 1 shows the heat flux observations of Lachenbruch & Sass's table 1, for the latitude range 35–40° N, where the deformation is predominantly transcurrent.

As Lachenbruch & Sass (1980) point out, the heat flux anomaly is far too broad to be explicable by dissipative heating on the seismogenic portion (upper 10–15 km) of the San Andreas fault. Nevertheless, the relatively sharp decrease in heat flux at the edges of the anomaly argues for a shallow source. This line of reasoning led Lachenbruch & Sass (1980) to consider dissipative heating within a broad shear zone as one plausible cause for the heat flux anomaly.

Lachenbruch & Sass's mechanical analysis of the heat flux anomaly postulated distributed shear on horizontal planes within a weak zone which decoupled, to a greater or lesser degree, the upper crust in the neighbourhood of the fault from the surrounding plates. An alternative view, taken by Molnar (1991) and in the next section, is to take upper crust as being *well* coupled to a relatively stronger viscous lower lithosphere, whose deformation it follows. According to that view, the deformation accommodating the relative motion of the Pacific and North America plates is expected to be spread out over some tens or hundreds of kilometres, in a fashion dictated by the rheology of the lithosphere as a whole (see discussion surrounding equation (4) below), and dissipative heating near the San Andreas fault is the result of shear on vertical planes in the strong lower lithosphere.

Let us take the intensity of the heat flux anomaly in central California to be 30 mW m^{-2}, spread over about 80 km width (figure 1). If shear on vertical planes over this same width accommodates the relative motion of the Pacific and North America plates, then the shear strain rate is *ca.* 2×10^{-14} s^{-1}. In a steady-state situation, the anomalous surface heat flux produced by dissipation is the product of a force per unit length, that is the integral from the surface to the base of the lithosphere of the associated deviatoric stress, and the strain rate, and these values of strain rate and anomalous heat flux imply that the integral of the shear stress through the lithosphere in the southern Coast Ranges is 1.5×10^{12} N m^{-1}. To obtain rough estimates of the shear stress, let us assume that the strong portion of the lithosphere is only 15 (30) km thick, and that stresses elsewhere in the lithosphere are much lower. The average shear stress in the strong layer is, on this assumption, 100 (50) MPa.

Molnar (1991) made a similar calculation but assumed a width of 200 km, instead of 80 km, for the deforming zone. Accordingly, he deduced a proportionally larger

Figure 1

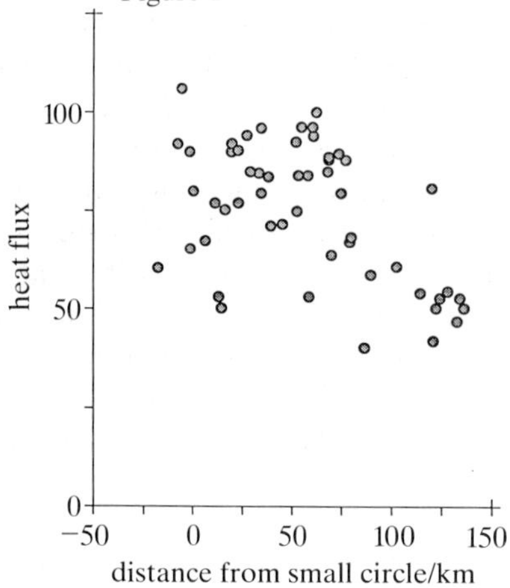

Figure 2

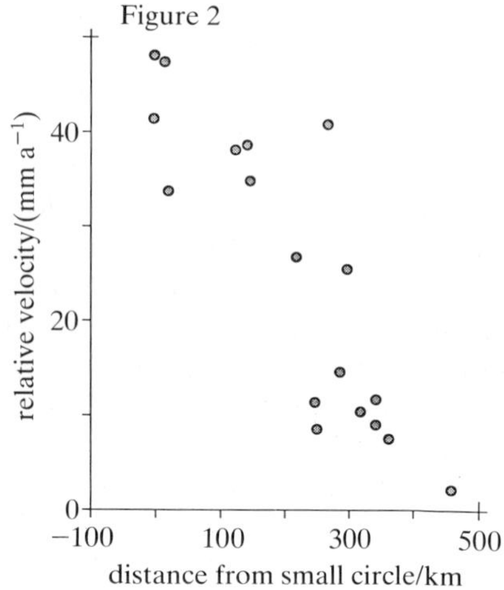

Figure 1. Observations of surface heat flow in California between 35° and 40° N, from Lachenbruch & Sass (1980, table 1). Observations from south of 35° N are excluded, where vertical advection of heat associated with crustal shortening in the Transverse Ranges, and with extension further south, presumably influences surface heat flux. Observations are plotted against distance from a small circle about the Euler pole describing the relative motion of the Pacific and North America plates (De Mets *et al.* 1990). The small circle passes through 38.1° N, 122.9° W (Point Reyes VLBI station).

Figure 2. Observed velocities of VLBI sites within California and Arizona. Velocities are observed in a frame that minimizes the velocities of VLBI sites within 'stable' North America (Ward 1990, tables 2 and 3, and fig. 3). The magnitudes of the velocities are plotted against distance from the same small circle used in figure 1. These data come predominantly from outside the region to which figure 1 refers, and do not provide an estimate of the shear strain rate within that region.

shear stress. Note that in southern California, where geodetic measurements suggest that the deformation is distributed over a width of 200 km or more (figure 2), there is no detectable heat flux anomaly associated with the San Andreas fault zone (Lachenbruch & Sass 1980, fig. 19). With the same shear stresses estimated above for central California, but with a shear strain rate three times smaller, the corresponding heat flux anomaly would be 10 mW m^{-2}. If, as seems likely, the shear stress diminishes with strain rate, this anomaly would be still lower.

The shear stresses estimated above for the lower lithosphere of the San Andreas fault system are similar in magnitude to the shear stresses estimated by Scholz *et al.* (1979) for mid-to-lower crustal levels near the Alpine fault of New Zealand. Thus, the apparently discrepant estimates of shear stress near major strike–slip faults that were cited at the beginning of this section may be reconciled by postulating that deformation of the uppermost crust of such regions is driven by tractions applied to its base by a stronger substrate that is deforming in a distributed fashion (Molnar 1991). This stronger substrate may lie in the lower crust or in the upper mantle, or in both.

3. Inferences of lithospheric strength from continuum calculations of continental deformation

Some success has been achieved in investigating the largest scale of continental deformation (hundreds to thousands of kilometres) by treating the continental lithosphere as a continuous medium (Bird & Piper 1980; England & McKenzie 1982, 1983; England *et al.* 1985; Houseman & England 1986; Vilotte *et al.* 1982, 1984, 1986). At the core of this approach is the recognition that most major zones of active continental deformation are much greater in their horizontal extents than they are thick (typically they are several hundreds to thousands of kilometres wide, whereas the lithosphere is about 100 km thick). This configuration, coupled with the assumption that *shear* tractions on the base of the continental lithosphere are generally negligible compared with those within it, yields the result that the deformation of the continental lithosphere depends principally upon the *vertical averages* of the stresses acting within it and on its rheology. The observations of strain rate and of topography that are summarized below are made over a horizontal scale of several hundreds of kilometres. It must be remembered that they are, therefore, incapable of directly yielding information on the variation with depth of stress or rheology.

At its simplest the continuum approach treats the lithosphere as a power law fluid whose rheology is characterized by two dimensionless parameters only (England & McKenzie 1982, 1983). The first of these parameters is the exponent in a power law that describes the rheology of the lithosphere (equation (3)). The second parameter (called the Argand number by England & McKenzie (1982)) expresses the ratio of the stresses associated with isostatically compensated density contrasts within the lithosphere (referred to here as 'buoyancy forces') to the stresses required to deform the lithosphere at a characteristic strain rate. These two parameters influence the deformation in different ways. The influence of the power law exponent in the rheology is seen most clearly in the way that it determines the shapes of deforming zones (England & McKenzie 1982, 1983; England *et al.* 1985), whereas the Argand number principally influences the magnitude of crustal thickness contrasts that can be supported by a given boundary condition (England & McKenzie 1982, 1983; Houseman & England 1986; Vilotte *et al.* 1986).

(*a*) *Distribution of strain within deforming regions*

The representation of the vertical average of the rheology of continental lithosphere by a power law is partly a matter of convenience. A relationship between strain rate, $\dot{\epsilon}$, and deviatoric stress τ of the form,

$$\dot{\epsilon} \propto \tau^n, \tag{3}$$

can represent, with the variation of a single parameter, a wide range of flow laws, from that of a newtonian fluid with $n = 1$, to a purely plastic material as $n \rightarrow \infty$. Equation (3) is not merely convenient, however. Sonder & England (1986) showed it to be a close approximation to the vertical average of the rheology of lithosphere failing by slip on faults in its upper crust and by a combination of creep and brittle failure in the lower lithosphere, according to the strain rate and temperature there. (The success of this approximation derives from the ease with which one can obtain a straight line by plotting the logarithm of one quantity against the logarithm of another, not from any single fundamental physical process.) Sonder & England (1986) point out that the exponent n in equation (3) does not represent any individual deformation mechanism. Rather, it represents the vertical averaging of deformation involving slip on faults (for which stress is essentially independent of strain rate, and therefore $n \approx \infty$) with deformation by creep (commonly with $n \approx 3$).

If continental lithosphere shows deformation that is consistent with its vertically averaged rheology having $n \approx 3$, then it may be assumed that a relatively small fraction of the total strength of the lithosphere is supported by friction on faults, and that most of the strength resides in the creeping lower lithosphere; in contrast, values of n higher than 10 indicate that a substantial fraction of the strength may be supported by friction on faults (Sonder & England 1986).

If continental lithosphere does behave as a thin sheet of fluid that obeys equation (3), then, in the absence of buoyancy forces, deformation associated with a boundary of length, D, ought to die out exponentially with distance from that boundary over a characteristic distance that is proportional to D and is inversely proportional to the square root of n in the vertically averaged rheology (equation (3)) (England *et al.* 1985). Furthermore, the across-strike width of the deforming region should be four times smaller for a transcurrent boundary than for a compressional or extensional boundary of the same length. Because the aspect ratios of the major active zones of continental deformation are in qualitative agreement with this prediction (England & Jackson 1989), it seems worthwhile to try to extract, from the aspect ratios of actively deforming regions, estimates of the exponent, n, that characterize the vertically averaged rheology of the lithosphere of those regions. Because compressional or extensional deformation leads inevitably to changes in crustal thickness, and because lateral variation in crustal thickness is a major contributor to buoyancy forces within continental lithosphere, the discussion below is restricted to inferences of n from the aspect ratios of zones of transcurrent deformation, where buoyancy forces may be neglected. In particular, we concentrate upon the Pacific–North-America plate-boundary zone in California, upon the South Island of New Zealand, and upon the Pacific Northwest of North America.

The characteristic across-strike dimension, l, of a zone of transcurrent deformation, of along-strike length D, in lithosphere whose vertically averaged rheology is given by equation (3), is:

$$l \approx D/\pi\sqrt{n} \tag{4}$$

(Sonder *et al.* 1986).

Analysis of VLBI baseline lengths between stations on the Pacific and North America plates showed that the rate of displacement between these stations and stable North America increases smoothly with distance from the Pacific–North-America pole over a zone that is about 300 km wide (figure 2) (Sauber 1989; Ward 1990). It is clear from each study that velocities within this deforming region are nearly parallel to the overall Pacific–North-America relative motion. The distance over which the relative velocity falls from the full Pacific–North-America velocity of about 50 mm a^{-1} to $1/e$ of that value is about 250 km (figure 2).

It is, of course, very difficult to distinguish, on the basis of geodetic measurements spanning a decade, or even a century, between elastic strain and permanent deformation. Geodetic observations cannot, therefore, be taken as directly relating to the deformation of the lower lithosphere. In southern California, however, the uniform-sense rotation of crustal blocks shown by palaeomagnetic data (Luyendyk *et al.* 1985) suggests that the region underwent, since middle Miocene time, distributed permanent deformation over approximately the same lengthscale as is shown by the geodetic observations of strain. McKenzie & Jackson (1983) pointed out that such rotations of crustal blocks could be explained if the upper crust were regarded as a set of small rigid blocks floating on a viscous substrate whose deformation they followed. In particular the rate of rotation, $\dot{\theta}$, of roughly circular blocks about a vertical axis would be:

$$\dot{\theta} = \tfrac{1}{2}(\partial u_x/\partial y - \partial u_y/\partial x), \tag{5}$$

where u_x and u_y are the x- and y-components of the velocity, and z is vertical. If the geodetic observations of figure 2 are taken to yield a gradient of velocity, say $\partial u_x/\partial y$ with x parallel to Pacific–North-America relative motion, and y perpendicular, then the rate of rotation of crustal blocks perfectly coupled to a fluid with such a velocity gradient would, at the boundary, be:

$$\dot{\theta} \approx \tfrac{1}{2} \times 55 \text{ mm a}^{-1}/250 \text{ km}, \tag{6}$$

or about 6° per million years, dropping to $1/e$ of that value at about 250 km from the boundary.

Average rotation rates about vertical axes determined from late Neogene rocks in southern California are a few degrees per million years (see Jackson & Molnar 1990, table 2). These observations support qualitatively the notion that the deformation of the upper crust of California follows passively the velocity gradient field of a continuous substrate. Sonder *et al.* (1986) also reached this conclusion by using a continuum calculation of the deformation associated with the past 29 Ma plate interactions in the region to analyse the palaeomagnetic data. Jackson & Molnar (1990) show that the active faulting and block rotation of the western Transverse Ranges are consistent with the motion of crustal blocks driven by tractions applied to their bases by a viscous substrate.

Making use of equation (4) with $l = 250$ km and taking a plate-boundary length of 1200 km (Salton Sea to Cape Mendocino), yields an estimate of $n \approx 3$ for the vertically averaged rheology of the lithosphere in California.

The Pacific Northwest of North America has experienced transcurrent deformation as a result of the relative motion of the North America and Juan de Fuca plates. Observations of palaeomagnetically determined rotations of rocks of the Columbia River Basalt Group allowed England & Wells (1991) to estimate the across-strike lengthscale of that deformation. The rotations die out to about $1/e$ of their values near the coast over a distance of about 150 km. The along-strike length of the plate boundary is about 1200 km which, from equation (4), gives an estimate of $n \approx 4.5$.

Determinations of strain rate in the South Island of New Zealand, based on retriangulation spanning about a century, and summarized and interpreted by Walcott (1984), show that the transcurrent deformation associated with the relative motion of the Pacific and Australia plates is distributed through a zone approximately 150 km wide in which the shear strain rate diminishes with distance east of the Alpine fault. The along-strike length of this zone is about 800 km. These dimensions yield a ratio $D/l \approx 5.5$, implying $n \approx 3$ if, as for California, the geodetic shear strain reflects distributed deformation of the lower lithosphere and not elastic strain of the upper crust.

(*b*) *Topographic contrasts supported by continental lithosphere*

The distribution of strain rates within active compressional belts may be used to infer levels of vertically averaged deviatoric stress in lithosphere of these regions. The presence of large plateaux in the Andes, Iran and Tibet implies that the stresses required to drive deformation of the lithosphere in these regions are in approximate balance with the buoyancy forces associated with the thickened crust underlying the plateaux (Molnar & Tapponnier 1978; Dalmayrac & Molnar 1981; England & McKenzie 1983; Houseman & England 1986; Vilotte *et al.* 1986; England & Houseman 1988, fig. 3; Molnar & Lyon-Caen 1988). In particular, both the Andes and Tibet exhibit a transition in tectonic style that is strongly correlated with surface elevation. Each region is in an overall convergent setting, and is surrounded by thrust faulting, yet in its highest part, each is undergoing normal faulting and crustal extension. The transition between extensional and thrust faulting earthquakes in the region of Tibet occurs at an elevation of about 4 km (England & Molnar, unpublished data), and in the Andes Quaternary normal faulting (almost seismically inactive during the instrumental period) is found above about 3–4 km, although the transition is less clear-cut than in Tibet (Dalmayrac & Molnar 1981; Sébrier *et al.* 1985).

These observations allow us to identify the surface height in each region at which the vertically averaged vertical stress acting on horizontal planes, $\bar{\sigma}_{zz}$, is equal to the vertically averaged horizontal stress acting on vertical planes, $\bar{\sigma}_{xx}$ (Dalmayrac & Molnar 1981). Because $\bar{\sigma}_{xx}$ varies slowly with distance, estimating $\bar{\sigma}_{zz}$ at the transition between normal and thrust faulting yields an estimate of the horizontal driving stress supporting the entire mountain range (Dalmayrac & Molnar 1981).

Unfortunately, determining $\bar{\sigma}_{zz}$ associated with a given surface height is not a simple matter. England & Houseman (1989, Appendix A), show that the magnitude of $\bar{\sigma}_{zz}$ is so dependent upon assumptions about the density distribution within the lithosphere, that a quantitative estimate of $\bar{\sigma}_{zz}$ based upon surface height alone is impossible. Lateral variations in $\bar{\sigma}_{zz}$ are directly related to variations in geoid height (Turcotte & Schubert 1982), but the geoid height is generally poorly determined over mountainous regions.

Molnar & Lyon-Caen (1988) estimate that the difference between $\bar{\sigma}_{zz}$ for the 5 km high Tibetan plateau and its surrounding lowlands is 70–80 MPa, or $7\text{–}8 \times 10^{12}$ N m^{-1} over a lithosphere 100 km thick. This difference is proportional to the difference between the squares of the crustal thicknesses of the two regions (England & McKenzie 1982). Let us take 35 km to be the thickness of continental crust whose surface lies at sea level, and 65 km for the thickness of crust whose surface lies at 5 km above sea level. The difference $\bar{\sigma}_{zz} - \bar{\sigma}_{xx}$ for crust whose surface height is at 4 km above sea level is, then, $5\text{–}6 \times 10^{12}$ N m^{-1}. The transition to normal faulting in the Andes takes place at a somewhat lower elevation, about 3 km (Sébrier *et al.* 1985; Mercier 1991). By the same kind of calculation, the driving force supporting the Andes at present is about $3\text{–}4 \times 10^{12}$ N m^{-1}. These driving forces are too large to be attributed to the driving force from the mid-ocean ridges, but descending and negatively buoyant lithosphere like the slab beneath the Andes can add appreciably to the driving force (Bott 1990; Bott *et al.* 1989; Molnar 1988*b*; Lyon-Caen & Molnar 1985).

In their present tectonic régimes the plateaux show an approximate balance between buoyancy forces arising from crustal thickness contrasts and the driving forces applied to their edges by the convergence of the blocks surrounding them. In the past, however, the plateaux were lower, and the driving forces were resisted mainly by the shear stresses required to deform the lithosphere (England & Houseman 1988, fig. 3; England & McKenzie 1983; Houseman & England 1986). Furthermore, each plateau is bordered by lower-lying regions that are now shortening in response to these same driving forces. Consequently, we may presume that the continental lithosphere of these plateaux, and their surroundings, can support driving forces of about $3–6 \times 10^{12}$ N m^{-1} at geological strain rates.

4. Conclusion

The heat fluxes observed, and the temperatures inferred from metamorphic petrology, in major thrust zones imply that the shear stresses in such zones are generally between 50 and 150 MPa (§2*a*). Shear stresses of this magnitude persist to depths of 40 km or greater, and can contribute $2–4 \times 10^{12}$ N m^{-1} (or more if such stresses are maintained at depths greater than 40 km) to the support of mountain ranges bordering these thrust zones. The relations between the style of active faulting and surface elevation in the Andes and Tibet (§3*b*) imply that the continental lithosphere of these regions requires a driving force of $3–6 \times 10^{12}$ N m^{-1} to overcome resistance to deformation. These driving forces are equivalent to vertically-averaged shear stresses of 15–30 MPa over a lithosphere 100 km thick, but if supported by, a layer for example, 30 km thick within the lithosphere, they would require shear stresses of 50–100 MPa.

The aspect ratios of the major active zones of transcurrent deformation (§3*a*) suggest that the vertically averaged rheology of the lithosphere is described by a power law (equation (3)) with $n \approx 3$. This observation implies that deformation by high-temperature creep is more important in determining the strength of the lithosphere of these regions than is friction on faults. The clockwise rotations recorded in Neogene rocks of western North America show that the lithosphere of the region has undergone distributed deformation in response to strike–slip boundary conditions, in a fashion consistent with the deformation of a viscous fluid (§3*a*). This implication is supported, but not proved, by the geodetic determinations of strain summarized in figure 2. Estimates of the shear stress acting below the seismogenic upper crust of the San Andreas fault system, and in amphibolite-grade rocks exposed near the Alpine fault of New Zealand (§2*a*), lie in the range 50–100 MPa. Note that these estimates are not at variance with the conclusions of Mount & Suppe (1987) and Zoback *et al.* (1987) that the shear stress in the upper crust on the San Andreas fault is low.

Several other lines of evidence also suggest that the upper crust does not control the regional distribution of deformation in tectonically active parts of the continents. The distributions of strain and strain rate in central and southern Asia agree well with those calculated assuming that the lithosphere behaves as a continuous, power-law, fluid (England & Houseman 1986). Even such apparently dominant features as the major strike–slip faults in eastern Tibet are probably rotating, in response to the continuous deformation of their substrate (England & Molnar 1990). Similarly, the distributions of active strain rate and total extensional strain in the Aegean region appear to be consistent with those of a temperature-dependent fluid such as olivine, rather than with the strain of lithosphere controlled by friction on faults (Sonder & England 1989).

The observations of distributed strain in active belts, and the inferences of shear stress magnitudes within these belts, support the idea that the uppermost, seismically active, part of the continental lithosphere is generally much weaker than a layer beneath it that deforms in a continuous fashion. This layer probably contains the uppermost mantle, and may include the lower crust. It seems likely that the motions of, and the stresses within, the upper crust reflect the response of this heterogeneous, anisotropic, and discontinuous layer to tractions applied to its base by distributed deformation of the stronger lower lithosphere. Information on the state of stress in the lower lithosphere is sparse and indirect. Perhaps the most valuable measurement that could be made to constrain better the state of stress in deforming belts would be an improved determination of the gravity field of the continents.

We thank M. H. P. Bott for a helpful review. This work was supported by NERC grant GR3/7032 and NASA grant NAG5-795.

References

Barton, C. M. & England, P. C. 1979 Shear heating at the Olympos (Greece) thrust and the deformation properties of carbonates at geological strain rates. *Geol. Soc. Am. Bull.* **90**, 483–492.

Bird, P. & Piper, K. 1980 Plane stress finite element models of tectonic flow in southern California. *Phys. Earth planet. Inter.* **21**, 158–175.

Bott, M. H. P. 1990 Stress distribution and plate boundary force associated with collision mountain ranges. *Tectonophys.* **182**, 193–209.

Bott, M. H. P., Waghorn, G. D. & Whittaker, A. 1989 Plate boundary forces at subduction zones and trench-arc compression. *Tectonophys.* **170**, 1–15.

Brereton, R. & Müller, B. 1991 European stress: contributions from borehole breakouts. *Phil. Trans. R. Soc. Lond.* A **337**, 165–179. (This volume.)

Brune, J. N., Henyey, T. L. & Roy, R. F. 1969 Heat flow, stress, and rate of slip on the San Andreas fault, California. *J. geophys. Res.* **74**, 3821–3827.

Chen, W.-P. & Molnar, P. 1983 Focal depths of intra-continental and intraplate earthquakes and their implications for the thermal and mechanical properties of the lithosphere. *J. geophys. Res.* **88**, 4183–4214.

Dalmayrac, B. & Molnar, P. 1981 Parallel normal and thrust faulting in Peru, and constraints on the state of stress. *Earth planet. Sci. Lett.* **55**, 473–481.

De Mets, C., Gordon, R. G., Argus, D. F. & Stein, S. 1990 Current plate motions. *Geophys. J. Int.* **101**, 425–478.

Ekström, G. & England, P. C. 1989 Seismic strain rates in regions of distributed continental deformation. *J. geophys. Res.* **94**, 10231–10258.

England, P. C. & Houseman, G. A. 1986 Finite strain calculations of continental deformation. 2. Comparison with the India-Asia collision zone. *J. geophys. Res.* **91**, 3664–3676.

England, P. C. & Houseman, G. A. 1988 The mechanics of the Tibetan Plateau. *Phil. Trans. R. Soc. Lond.* **326**, 301–319.

England, P. C. & Houseman, G. A. 1989 Extension during continental convergence with application to the Tibetan plateau. *J. geophys. Res.* **94**, 17561–17579.

England, P. C. & Jackson, J. A. 1989 Active deformation of the Continents. *A. Rev. Earth Sci.* **17**, 197–226.

England, P. C. & McKenzie, D. P. 1982 A thin viscous sheet model for continental deformation. *Geophys. Jl R. astr. Soc.* **70**, 295–321.

England, P. C. & McKenzie, D. P. 1983 Correction to: A thin viscous sheet model for continental deformation. *Geophys. Jl R. astr. Soc.* **73**, 523–532.

England, P. & Molnar, P. 1990 Surface uplift, uplift of rocks, and the exhumation of rocks. *Geology* **18**, 1173–1177.

England, P. & Wells, R. 1991 Neogene rotations and quasicontinuous deformation of the Pacific Northwest Continental Margin. *Geology*. (In the press.)

England, P. C., Houseman, G. A. & Sonder, L. J. 1985 Lengthscales for continental deformation in convergent, divergent and strike–slip environments: analytical and approximate solutions for a thin viscous sheet model. *J. geophys. Res.* **90**, 3551–3557.

England, P., Le Fort, P., Molnar, P. & Pêcher, A. 1991 Heat sources for Tertiary metamorphism and anatexis in the Annapurna–Manaslu region, central Nepal. *J. geophys. Res.* (In the press.)

Fuchs, K., Zoback, M. D. & Rummel, F. 1991 A stress measurement profile to mid-crustal depth in KTB scientific drilling project, southeastern Germany: scientific questions and technological challenges. *Phil. Trans. R. Soc. Lond.* A **337**, 25–28. (This volume.)

Graham, C. M. & England, P. C. 1976 Thermal regimes and regional metamorphism in the vicinity of overthrust faults: an example of shear heating and inverted metamorphic zonation from southern California. *Earth planet. Sci. Lett.* **31**, 142–152.

Houseman, G. A. & England, P. C. 1986 Finite strain calculations of continental deformation. 1. Method and general results for convergent zones. *J. Geophys. Res.* **91**, 3651–3663.

Hubbard, M. S. 1989 Thermo-barometric constraints on the thermal history of the Main Central Thrust zone and the Tibetan slab, Eastern Nepal, Himalaya. *J. Metam. Petrol.* **7**, 19–30.

Jackson, J. A. & McKenzie, D. P. 1984 Active tectonics of the Alpine–Himalayan belt between western Turkey and Pakistan. *Geophys. Jl R. astr. Soc.* **77**, 185–264.

Jackson, J. A. & McKenzie, D. P. 1988 The relationship between plate motions and seismic moment tensors, and the rates of active deformation in the Mediterranean and the Middle East. *Geophys. Jl R. astr. Soc.* **93**, 45–73.

Jackson, J. A. & Molnar, P. 1990 Active faulting and block rotation in the western Transverse Ranges, California. *J. geophys. Res.* **95**, 22073–22087.

Kao, H. & Chen, W.-P. 1991 Earthquakes along the Ryukyu–Kyushu arc: strain segmentation, lateral compression and the thermo-mechanical state of the plate interface. *J. geophys. Res.* (Submitted.)

Kostrov, B. V. 1974 Seismic moment and energy of earthquakes, and seismic flow of rock. *Izv. akad. Nauk SSSR:* Phys. Solid Earth **1**, 23–44.

Lachenbruch, A. H. & Sass, J. H. 1980 Heat flow and energetics of the San Andreas fault zone. *J. geophys. Res.* **85**, 6185–6222.

Le Fort, P. 1975 Himalayas: the collided range. Present knowledge of the continental arc. *Am. J. Sci.* A **275**, 1–44.

Luyendyk, B. P., Kamerling, M. J., Terres, R. R. & Hornafius, J. S. 1985 Simple shear of southern California during Neogene time suggested by Palaeomagnetic declinations. *J. geophys. Res.* **90**, 12454–12466.

Lyon-Caen, H. & Molnar, P. 1985 Gravity anomalies, flexure of the Indian plate, and the structure, support and evolution of the Himalay and Ganges Basin. *Tectonics* **4**, 513–538.

McKenzie, D. P. 1978 Active tectonics of the Alpine–Himalayan belt: the Aegean Sea and surrounding regions. *Geophys. Jl R. astr. Soc.* **55**, 217–254.

McKenzie, D. P. & Jackson, J. A. 1983 The relationship between strain rates, crustal thickening, palaeomagnetism, finite strain and fault movements within a deforming zone. *Earth planet. Sci. Lett.* **65**, 182–202.

Molnar, P. 1988*a* A review of the geophysical constraints on the deep structure of the Tibetan plateau, the Himalaya and the Karakoram, and their tectonic implications. *Phil. Trans. R. Soc. Lond.* **326**, 33–88.

Molnar, P. 1988*b* Continental tectonics in the aftermath of plate tectonics. *Nature, Lond.* **335**, 131–137.

Molnar, P. 1991 Brace–Goetze strength profiles, the partitioning of strike-slip and thrust faulting at zones of oblique convergence and the stress–heat flow paradox of the San Andreas fault. In *Fault mechanics and transport properties of rocks: a Festschrift in honor of W. F. Brace* (ed. B. Evans & T.-F. Wong). London: Academic Press.

Molnar, P. & Deng, Q.-D. 1984 Faulting associated with large earthquakes and the average rate of deformation in central and eastern Asia. *J. geophys. Res.* **89**, 6203–6228.

Molnar, P. & England, P. 1990 Temperatures, heat flux, and frictional stress near major thrust faults. *J. geophys. Res.* **95**, 4833–4856.

Molnar, P. & Lyon-Caen, H. 1988 Some simple physical aspects of the support, structure and evolution of mountain belts. In *Processes in continental and lithospheric deformation* (ed. S. P. Clark Jr). Geol. Soc. Am. Spec. Pap. 218.

Molnar, P. & Lyon-Caen, H. 1989 Fault plane solutions of earthquakes and active tectonics of the Tibetan plateau and its margins. *Geophys. J. Int.* **99**, 123–153.

Molnar, P. & Tapponnier, P. 1975 Cenozoic tectonics of Asia: effects of a continental collision. *Science, Wash.* **189**, 419–426.

Molnar, P. & Tapponnier, P. 1978 Active tectonics of Tibet. *J. geophys. Res.* **83**, 5361–5375.

Mount, V. S. & Suppe, J. 1987 State of stress near the San Andreas fault: implications for wrench tectonics. *Geology* **15**, 1143–1146.

Sauber, J. 1989 Geodetic measurement of deformation in California. Ph.D. thesis, MIT, Cambridge.

Scholz, C. H. 1980 Shear heating and the state of stress on faults. *J. geophys. Res.* **85**, 6174–6184.

Scholz, C. H., Bevan, J. & Hanks, T. C. 1979 Frictional metamorphism, argon depletion, heat flow, and stress on the Alpine fault, New Zealand. *J. geophys. Res.* **84**, 6770–6882.

Sébrier, M., Mercier, J. L., Mégard, F., Laubacher, G. & Carey-Gailhardis, E. 1985 Quaternary normal and reverse faulting and the state of stress in the central Andes of Peru. *Tectonics* **4**, 739–780.

Sonder, L. J. & England, P. C. 1986 Vertical averages of rheology of the continental lithosphere: relation to thin sheet parameters. *Earth planet. Sci. Lett.* **77**, 81–90.

Sonder, L. J. & England, P. C. 1989 Effects of a temperature-dependent rheology on large scale continental extension. *J. geophys. Res.* **94**, 7603–7619.

Sonder, L. J., England, P. C. & Houseman, G. A. 1986 Continuum calculations of continental deformation in transcurrent environments. *J. geophys. Res.* **91**, 4797–4810.

Turcotte, D. L. & Schubert, G. 1982 *Geodynamics.* New York: Wiley.

Vilotte, J. P., Daignières, M. & Madariaga, R. 1982 Numerical modeling of intraplate deformation: Simple mechanical models of continental collision. *J. geophys. Res.* **87**, 10709–10728.

Vilotte, J. P., Daignières, Madariaga, M., R. & Zienkiewicz, O. C. 1984 The role of a heterogeneous inclusion during continental collision. *Phys. Earth planet. Int.* **36**, 236–259.

Vilotte, J. P., Madariaga, R., Daigneières, M. & Zienkiewicz, O. C. 1986 Numerical study of continental collision: influence of buoyancy forces and a stiff inclusion. *Geophys. Jl R. astr. Soc.* **84**, 279–310.

Walcott, R. I. 1984 The kinematics of the plate boundary through New Zealand: a comparison of the short and long term deformations. *Geophys. Jl R. astr. Soc.* **79**, 613–633.

Ward, S. N. 1990 Pacific–North America plate motions: new results from very long baseline interferometry. *J. geophys. Res.* **95**, 21965–21983.

Zoback, M. D. 1991 State of stress and crustal deformation along weak transform faults. *Phil. Trans. R. Soc. Lond.* A **337**, 141–150. (This volume.)

Zoback, M. L. & Magee, M. 1991 Stress magnitudes in the crust: constraints from stress orientation and relative magnitude data. *Phil. Trans. R. Soc. Lond.* A **337**, 181–194. (This volume.)

Zoback, M. D. *et al.* 1987 New evidence on the state of stress of the San Andreas fault system. *Science, Wash.* **238**, 1105–1111.

European stress: contributions from borehole breakouts

By Robin Brereton[1] and Birgit Müller[2]

[1] *Regional Geophysics Group, British Geological Survey, Nicker Hill, Keyworth, Nottingham NG12 5GG, U.K.*

[2] *Geophysikalisches Institut, Universität Fridericiana Karlsruhe, Hertzstrasse 16 7500 Karlsruhe-West, F.R.G.*

Before about 1986 European measurements of rock stresses were sparse. The development of the borehole breakout method has facilitated a dramatic increase in the geographical distribution of rock stress orientations. These are derived from data collected routinely for other purposes by the oil, coal and gas industries over many years. Borehole breakouts can be interpreted in terms of the orientation of the stress field in boreholes as a function of depth, with the findings related to known stratigraphical boundaries, tectonic structures, fault and fracture occurrences, and to other geological and geophysical stress indicators. Geographical and depth distributions of stress on local and regional scales are beginning to provide the necessary base from which broad interpretive stress field models can be constructed.

1. Introduction

A growing awareness of the importance or rock stress measurements to the understanding of broad tectonic and deep geological processes has led to an increased effort of collating and analysing stress data throughout Europe. Until the mid-1980s the number of rock stress determinations was small and relied primarily on earthquake focal mechanisms or on relatively few near surface (less than 50 m) measurements. There is still a great paucity of data but, more importantly, our understanding of the geological origins of rock stress and of the interplay between stress, structure, tectonic evolution and other geological attributes on a variety of scales remains limited. One of the advantages of incorporating European rock stress determinations into a coordinated database is that subsequent regional and local stress maps, and the interpretations based upon them, will be subject to the rigours of the wider quality control procedure established for the World Stress Map.

Studies in recent years have demonstrated that borehole breakouts can be used as a reliable indicator of the orientation of principal horizontal *in situ* rock stresses (Zoback *et al.* 1989). The advent of stress measurements from breakouts and the ready availability of primary data from existing boreholes has permitted a rapid expansion of the European Stress Map database from which both regional and local stress fields can be derived.

For much of northwestern Europe there prevails a broadly consistent NW–SE maximum compressive stress orientation ($\sigma_{\mathrm{H,max}}$) from the Alps to northern Britain. This arises from plate movements and can be attributed to forces associated with the interaction between stress fields created by the east to southeasterly separation of Europe from North America and the north to northeasterly movement between the African plate and Europe.

Phil. Trans. R. Soc. Lond. A (1991) **337**, 165–179
Printed in Great Britain

On a more pragmatic level, the energy industries are beginning to acknowledge the importance of stress field information as an aid to maximizing the economic development and exploitation of hydrocarbon, coal and geothermal resources. Low permeability, or 'tight', gas- and oil-bearing formations often require massive hydraulic fracture treatments to achieve economic rates of production. Hydraulically induced fractures will preferentially propagate in a direction parallel to the maximum stress orientation. In general, it is disadvantageous for zones of induced fracturing between adjacent wells to interfere with one another and thereby provide hydraulic short circuits. Local stress field information will allow optimum well field arrays to be designed to avoid this potential problem. On the other hand, stress field information will allow the prediction of whether off-target wells can be connected to a reservoir by fracturing or will enable the optimum location of relief wells to be drilled in blowout situations (Bell & Babcock 1986).

Hot dry rock geothermal energy depends upon the concept that water circulated between a couplet of boreholes will exchange heat with the rock and provide steam for electricity generation (Downing & Gray 1986). In this case, the water needs an efficient transmission path between the boreholes, usually provided by fracturing. A knowledge of the local stress field is required to predict the extent and direction of the induced fractures so that the borehole couplet can be properly oriented.

Rock stress orientations can have a significant influence on the economics of mining. Jackson *et al.* (1989) have shown that rectangular section coal mine roadways driven within 30° of the minimum horizontal stress direction can suffer marked deformation and support problems. Roadways at right angles to the minimum stress direction stood up well. Gale & Blackwood (1987) had arrived at comparable results after a series of three-dimensional analyses to determine stress distributions around a mine face when the roadway intersects the *in situ* stress field at different angles. Other large underground excavations such as electricity generator turbine halls and radioactive waste disposal facilities also require a comprehensive knowledge of the local stress field, preferably including relative stress magnitudes to allow the estimation of the likelihood of fault reactivation.

These examples of the scientific and commercial applicability of rock stress derivations serve to emphasize the importance of collating available data in the European Stress Map.

2. Scales of measurement

A complete characterization of the state of stress in a rock mass requires a three-dimensional description of both the stress orientation and stress magnitude. There is, however, no absolute method of measuring *in situ* rock stress against which other methods can be calibrated. This is in part because of the differences in scale over which the various measurement methods apply and the interplay between the forces at work within a rock mass; from the intergranular scale, through layered sediments, jointed blocks, plutons, major faults, and plates at the continental scale. Hyett (1990) visualized the state of stress in a rock mass as being assembled from self-equilibrating elements, stored within equilibrium volumes, superposed on one another over the complete spectrum of geological scales. Far-field stresses will be represented by an integration of all the near-field components. For fractured rock masses, the dimensions of the equilibrium volumes are controlled by fracture geometry; length, spacing, frequency of intersection, etc. In general, only that component of the residual stress, stored within equilibrium volumes greater than that of the measurement scale, will be sampled. When a rock mass is divided into units smaller than the equilibrium volume, part of that stress may be released. This may result in the determination of different stress tensors at different scales of measurement.

Stress measurement methods which work on hand-size samples, such as differential strain analysis, anelastic strain recovery or overcoring, will primarily relate to the intergranular and microfracturing scale. Earthquake focal mechanisms on the other hand, apply to scales of hundreds of metres to kilometres. Borehole methods, such as breakouts and hydrofracturing, are influenced by stresses stored on the more geologically representative layered scale and enable measurements from the near surface to depths of several kilometres.

Because of the influence of scale upon results derived from the various stress measurement methods, it is important to not only record a full range of results in databases such as the European Stress Map, but also to exercise caution when interpreting the data in geological terms. The scale of the interpretation, whether it be local, regional or continental, should take account of the scale of the measurements used to influence that interpretation.

3. Regional stress fields

The European subset of the World Stress Map database contains close to 1500 entries (Müller *et al.* 1991). These data are classified, according to reliability as tectonic stress indicators, into quality categories A to E, where A to C are regarded as very reliable. The database contains previously existing compilations including the Fennoscandian Rock Stress Database (Stephansson *et al.* 1986), the Catalogue of Focal Mechanisms for European Earthquakes (Udias *et al.* 1989) and other, previously unpublished listings. The majority of the data (59%) are from focal mechanisms, followed by overcoring measurements (19%), borehole breakouts (14%), hydraulic fracturing (5%), and fault slip orientations, which fall under the category of geological stress indicators (3%).

The geographical and depth distributions of the data are variable: focal mechanisms concentrate in seismogenic zones and occur mostly at depths greater than 5 km; borehole breakout and hydraulic fracturing data are, as a rule, from sedimentary basins usually to depths less than 5 km. Overcoring and fault-slip data are the shallowest, with the exception of some measurements taken in tunnels and mines. Although these stress indicators monitor different scales of stress, in places where it is possible to make comparisons Müller *et al.* (1991) showed that significant differences in the orientations of horizontal stresses are not apparent.

From the data coverage displayed in figure 1, Europe can be subdivided into three major stress provinces.

1. Western Europe, broadly covering an area between 45° N to 55° N and 10° W to 17° E. The stress orientation is generally very uniform with a mean of $145 \pm 26°$. This homogeneous stress field is coincident with a thin to medium lithospheric thickness (50–90 km) and a medium to high heat flow (greater than 80 mW m^{-2}). A comparison of the observed horizontal stress directions with the trajectories of plate motion (Minster & Jordan 1978) indicates differences of about 17° compared with the absolute plate motion and about 2° compared with the relative plate motion between Eurasia and Africa. However, these relative motions

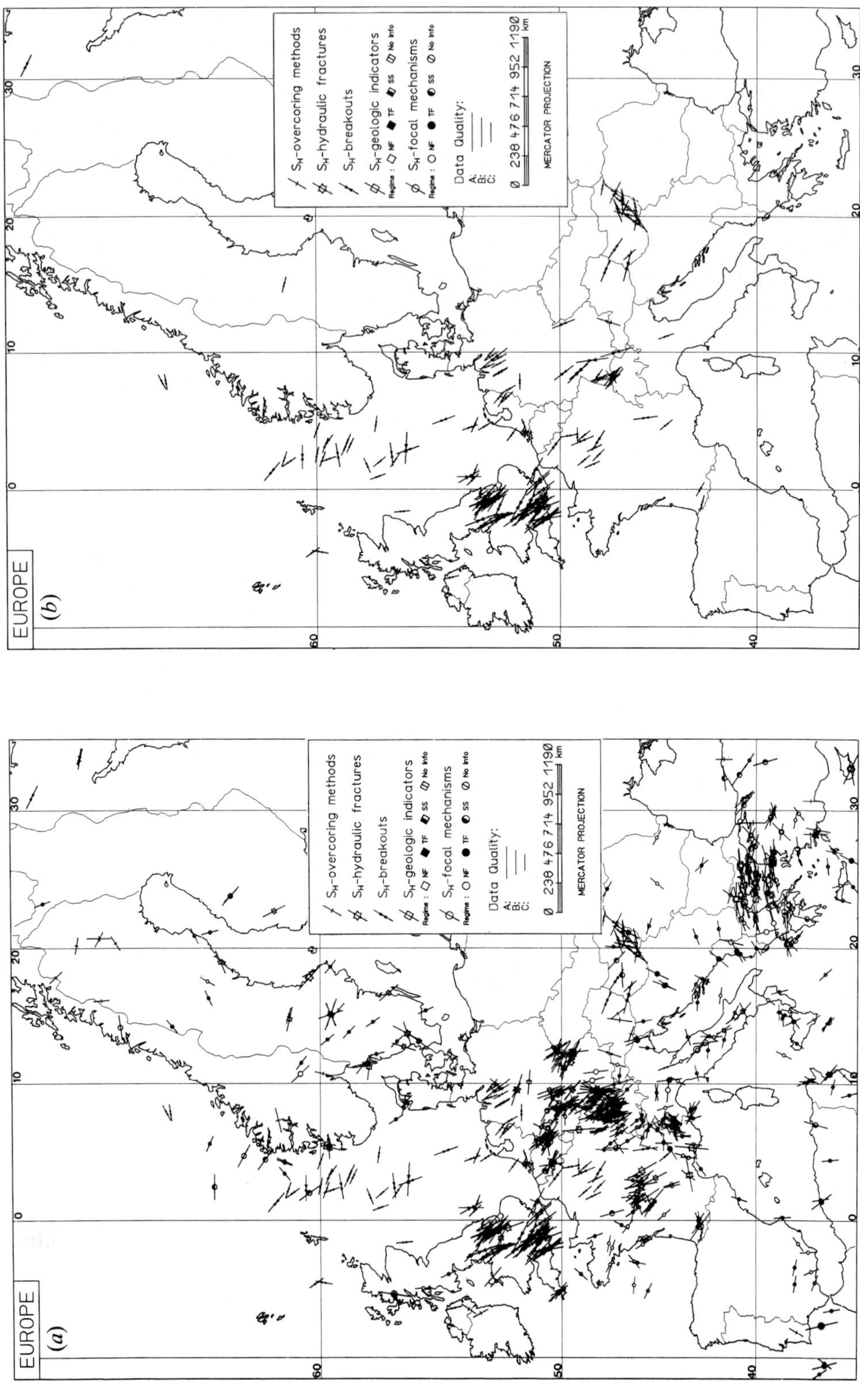

Figure 1. (*a*) European Stress Map. The symbols indicate the orientations of maximum horizontal compressive stresses for the quality categories 'A', 'B' and 'C'. The symbols represent the different types of stress indicator; longer symbols designate better qualities. (After Müller *et al.* 1991.) (*b*) All $\sigma_{\mathrm{H,max}}$ orientations deduced from borehole breakout observations.

depend very much upon how accurately the poles of plate motion can be determined, which is not very precise for Europe because of the very slow Eurasian plate movement.

2. Northern Europe covering latitudes greater than 55° N. The average orientation of $\sigma_{H,max}$ is $120\pm45°$, similar to, but not as consistent as, that for western Europe. Most of the area is characterized by a thick lithosphere (110–170 km) and a relatively low heat flow (less than 50 mW m^{-2}). Modelling work by Kusznir & Bott (1977) has demonstrated a reverse relationship between stress magnitudes, lithospheric thickness and geothermal gradient: the thinner the lithosphere and the larger the thermal gradient then the greater is the stress amplification in the upper crust. Stresses are therefore expected to be lower in magnitude for a thick, cold lithosphere and subject to greater influence by local perturbations. Regional stress field influences which also need to be taken into account are: the effect of plate-boundary forces (discussed by Stephansson 1988); post-glacial uplift (Bungum *et al.* 1991) and post-glacial flexure due to sediment loading along the continental margins (Stein *et al.* 1989); and large topographical gradients. The spreading of thick crust would create tensions perpendicular to the continental margin and compressions in adjacent oceans. Similarly, deglaciation would create localized tensions and compressions elsewhere. Swolfs & Savage (1985) have demonstrated good agreement between measured stresses and analytical solutions used to assess the influence of topography (see also Liu & Zoback 1991). The superposition of these influences upon lateral plate-boundary forces being transmitted from the mid-Atlantic Ridge may account for the diversity of stress orientations observed in northern Europe.

3. The Aegean and western Anatolia are characterized by N–S extension and a $\sigma_{H,max}$ orientation of $085\pm27°$. The tectonic setting of the eastern Mediterranean can be broadly described as a back-arc basin resulting from the northward movement of Africa relative to Eurasia. The basin is bounded by the Hellenic Trench to the west and south and by the North Anatolian fault to the north. Current lack of data prevents the easterly extrapolation of the stress field but it may well persist as far as the East Anatolian fault.

Further characteristics of the European stress field can be summarized as follows.

(*a*) The stress field is generally compressive with the intermediate principal stress being primarily vertical. This is suggested by the predominance of strike–slip faulting throughout much of Europe and accords with Anderson's (1951) hypotheses that, within the constraints of boundary conditions, the uppermost crustal stress field has one axis close to vertical: nearly 40% of the available data for which spatial orientations are given have one of the principal stress axes within 20° of vertical. Areas where the maximum stress orientation is vertical are the Lower Rhine Embayment, western France and the Aegean and western Anatolian region.

(*b*) The influence of regional-scale geological structures on stress field orientations is not ubiquitous. In the western Alps, for example, the horizontal stresses seem to align perpendicular to the Alpine front. On the other hand, the Rhinegraben rift system does not seem to perturb the regional-stress field. The graben can be subdivided into the Lower Rhine Embayment (striking NW) where normal faulting dominates, and the Upper Rhine Graben (mostly striking NNE) where strike–slip faulting is clear (Ahorner 1970, 1975). Even so, the stress orientation is constant within and across the graben system. Thus it seems that the relative geometrical juxtaposition of tectonic structures and stress orientations is responsible for the type of resultant deformation.

(*c*) On a broad regional scale, there seems to be no systematic variation of stress orientation with depth.

4. Borehole breakout contributions to the European stress field

Borehole breakout measurements have made a considerable contribution to our recent knowledge of the European state of stress. Breakouts describe the resultant elongation of a borehole cross section induced by the refraction and concentration of stress tensors near the free surface of a borehole wall. This stress concentration produces shear fracturing, microcrack propagation and associated spalling of material, in a direction parallel to the minimum stress orientation.

Breakouts are conventionally measured using information from the dipmeter logging tool. This is a four-arm pad device and will only provide *X–Y* caliper – and related pad resistivity information – at any particular depth in the borehole. The Borehole Televiewer (BHTV), on the other hand, is an acoustic imaging tool which is also able to give the full borehole cross-sectional shape. The BHTV can therefore be used to analyse breakouts in much more detail than can be obtained from the dipmeter. However, the BHTV is a relatively recent and novel device which is only run on special occasions, whereas the dipmeter is run almost always.

Breakout analyses can be performed not only by using caliper readings but also using the corresponding resistivity differences: referred to as caliper and resistivity eccentricity respectively (Brereton & Evans 1989; Brereton *et al.* 1990). Where microcrack propagation has allowed invasion of higher conductivity drilling fluids the formation resistivity will be reduced in the breakout azimuth. Resistivity eccentricity analysis can therefore be used to detect azimuthal zones of incipient spalling where rock material has not yet become dislodged. An effective method of displaying breakout data and enhancing dominant trends is the rose diagram. Results can also be weighted according to the relative size, or magnitude, of individual breakouts.

The principal advantages that the borehole breakout method has over other means of rock stress evaluation are (i) the large geographical distribution of appropriate data currently available; (ii) this type of data is routinely collected within the hydrocarbons and coal industries and as part of other deep geological investigations; (iii) it represents a very cost-effective means of providing detailed stress field measurements, always assuming the raw data have already been collected for other purposes; (iv) it helps to provide a tie between the shallow near surface stress measurements and the deeper focal mechanism stress indicators; (v) breakout data show the least variability in their regional scale mean orientation. In western Europe the mean $\sigma_{H,\max}$ orientation from breakout observations is $141 \pm 18°$.

Breakout analyses have been performed in nearly 200 boreholes drilled in a variety of European geological environments: sedimentary basins such as the Paris Basin, Pannonian Basin and the North German Plain; in the vicinity of the Rhinegraben and the Viking and Central graben structures of the North Sea; as well as into crystalline basement rocks (e.g. KTB and boreholes in the northern Alpine foreland). But, by far the most breakout measurements have been performed in Britain (Brereton & Evans 1989) where they form the majority of the stress map data. Here they are concentrated in the onshore exploration areas of the Midlands and central southern England. The overall picture is one of an average orientation of $\sigma_{H,\max}$ of $138 \pm 11°$(relative to geographic north) for the United Kingdom as a whole. In selected areas the results are a little more variable. In the Midlands there seems to be a bimodal trend at about 155° and 129°. The mean trend for southern England is 135 ± 10° while two subregions of the Hampshire Basin and Wytch Farm area have trends of 129° and 145° respectively. These directions concur with a system of northwest-striking mesofractures well developed throughout Upper Cretaceous and Palaeogene rocks in southern England and northern France (Bevan & Hancock 1986). The mean orientation for the whole country may reflect the averaging of a dominant trend of about 130° with a secondary trend of about 155°.

5. Local variations

Locally, lithospheric or regional stress fields can become significantly modified by the influence of both micro- and macroscale features. The most common are boundary effects associated with geological discontinuities or free surfaces, such as joints, faults, bedding planes, cavities, or the Earth's surface, which will cause stress trajectories to deviate.

Evidence suggests it is unlikely that a localized three-dimensional stress field in a complex geological setting would be uniform, homogeneous or simple. A spectrum of results is to be expected with features characteristic of both the micro- and macroscale influences but with dominant trends reflecting the larger regional-scale stress field.

The rotation of the horizontal stress field in the vicinity of the San Andreas fault is well documented (Zoback *et al.* 1989) and rotations relative to faulting along the Scotian Shelf can be inferred from Bell (1989). Breakout observations in coal exploration boreholes indicate that the free surface effect of the Earth's surface can extend to depths of 100–150 m and cause stress rotations away from the regional trend (K. R. Whitworth, personal communication). On the borehole scale gradual and discontinuous changes in breakout orientation have been associated with fractures and faults intersected by the hole and attributed to local perturbations of the stress field near the borehole due to its interaction with the fractured zones (Shamir & Zoback 1989).

Structurally controlled localized stress rotations were suggested by data presented by Springer (1987). Refraction of mean stress trajectories around a Precambrian basement high, producing a significant anticlockwise rotation relative to the surrounding regional trend, was ascribed by Bell & Lloyd (1989) to a contrast in the elastic properties between the crystalline basement and the overlying sediments. Bradshaw & Zoback (1988) argued that if a significant contrast in viscosity exists between two adjacent strata, such as a normally pressured sandstone and an underlying overpressured shale or evaporite, there will be a rotation of $\sigma_{\mathrm{H,max}}$ away from the vertical which will lead to the formation of low-angle normal faults.

Measured stress orientations, derived from borehole breakouts in wells around a salt dome in northern Germany (Schneider 1985), have been compared with the calculated principal stress directions using a finite element model: the shape of the salt dome was known from seismic investigations while representative material parameters were incorporated in the model (figure 2). There is clear agreement between the observations and calculated results. The principal stresses realign in the vicinity of the salt dome, whereas within the salt dome itself the near homogeneous stress is more representative of the far field stresses. The extent of the stress rotation is dependent upon the elastic parameter differences. Thus the salt behaves as a soft

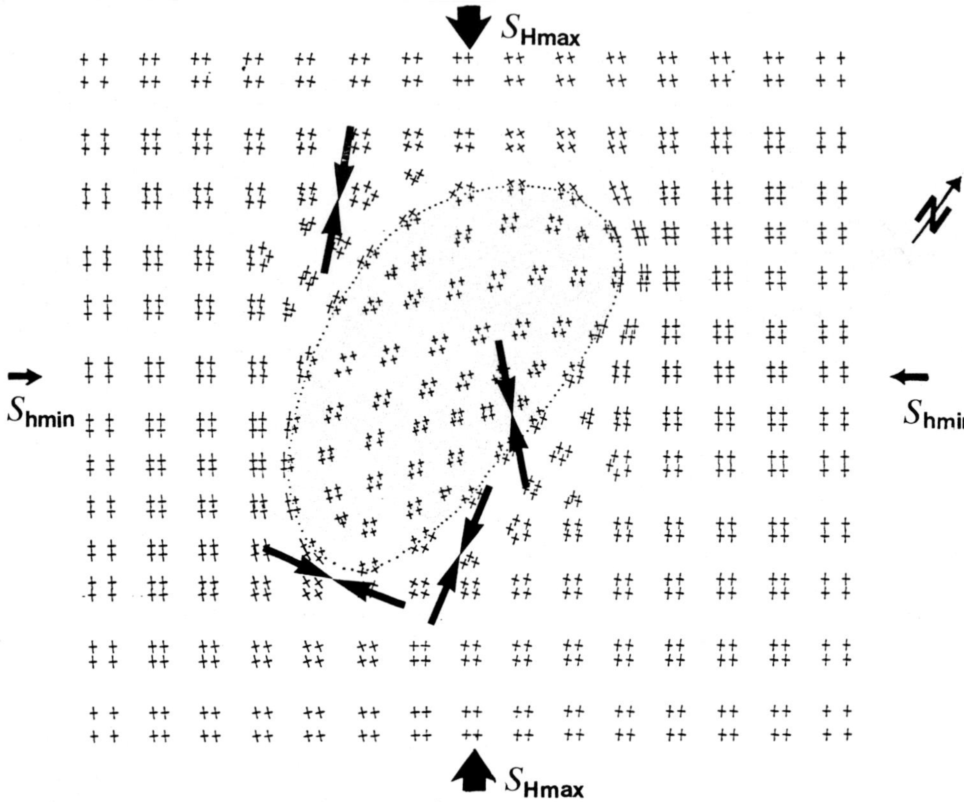

Figure 2. Observed orientations of principal horizontal stress around a salt dome in northern Germany compared with calculated stress orientations (crosses) using a finite-element model.

inclusion in the surrounding material (Goodier 1933). Stress rotations caused by bent dislocations and around elliptical inclusions have been comprehensively modelled by Ivins & Lyzenga (1986).

In contrast to these lateral and structural influences, the stress field of northern Switzerland exhibits changes primarily with depth (Clauss 1987; Müller *et al.* 1991). The block diagram in figure 3 summarizes current knowledge of the modern state of stress in an area between the Aare Massif and the folded Jura Mountains: (*a*) in the crystalline basement the stress orientation is about 140°, which is close to the western European mean; (*b*) in the Permocarboniferous Trough, situated to the north and east of the mountains, rock property contrasts seem to have caused the stress field to rotate to 160°; (*c*) south of the Jura Mountains the stress in the Molasse sediments is virtually N–S but changes, over a depth interval of less than 300 m, to 140°. Finite element modelling indicates that the Molasse stresses are caused by Alpine topographical effects, the Alps trend almost E–W in this area. This observed rotation of horizontal stress coincides with intensive plastic deformation of anhydrite layers, extending over a large horizontal distance, which is thought to vertically decouple the stress field.

Processes influencing localized variability have been studied near a British Coal colliery in the English Midlands (Brereton *et al.* 1990). The study area, of about 40 km^2, was chosen because of the large number of boreholes (over 30), and because the three-dimensional geology is well known. The tectonic evolution, from late

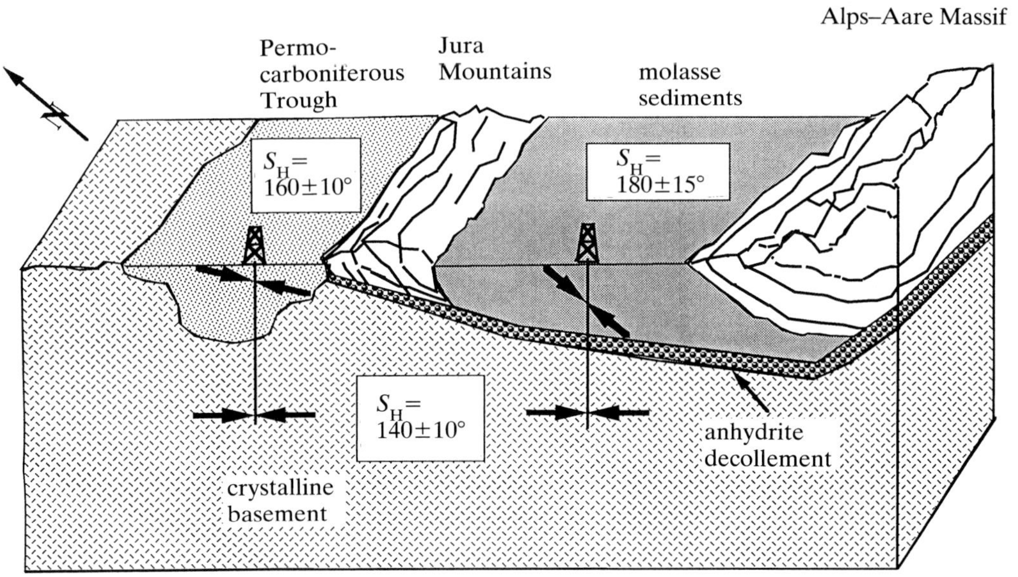

Figure 3. Sketch diagram of stress distribution in the folded Jura Mountains of northern Switzerland. The simplified block diagram shows stress orientation variations with depth due, in part, to an anhydrite decollement horizon between sediments of the Molasse Basin and the crystalline basement.

Carboniferous to early Permian times, resulted in the formation of horst and graben structures with segmented Coal Measures sequences. The grabens subsequently became depositional troughs and are sites of thick Permo-Triassic deposits. Two major faults, trending NE and N, abruptly displace the Coal Measures sequence with an associated downthrown-side thickening of the Permo-Triassic cover to the SE or E (of 130–200 m and 30–90 m respectively). Numerous northerly trending faults, with throws of only a few metres, are also known.

Fourteen boreholes were selected, clustered in two zones of contrasting geology. Cores exist for selected intervals only, but detailed lithostratigraphy from standard cuttings analysis or from using a combination of geophysical logs (British Coal ROCTEC: Halker *et al.* 1982) was available for all boreholes. The ROCTEC routines also provide a measure of relative rock strength from a fracture identification index (FIDX). Breakouts were determined from analyses of both caliper and resistivity data.

Almost all of the boreholes investigated displayed at least one conspicuous zone of breakouts. However, these zones often showed significant variations in depth, depth extent, host strata, orientation and magnitude of the associated breakouts. Even breakouts in boreholes a few hundred metres apart (e.g. Echills 1 and 2), do not correlate as consistently as might have been expected.

Despite the complexities of the breakout behaviour, several important observations could be made.

(*a*) Rose diagrams derived from the full depth extent of each borehole show two main trends of breakout elongation, ENE and NNW. The dominant ENE trend of minimum horizontal stress is consistent with breakout trends for Britain as a whole (Brereton & Evans 1989) and is also supported by overcoring studies performed in the local mine by British Coal. The secondary NNW trend was found primarily in a

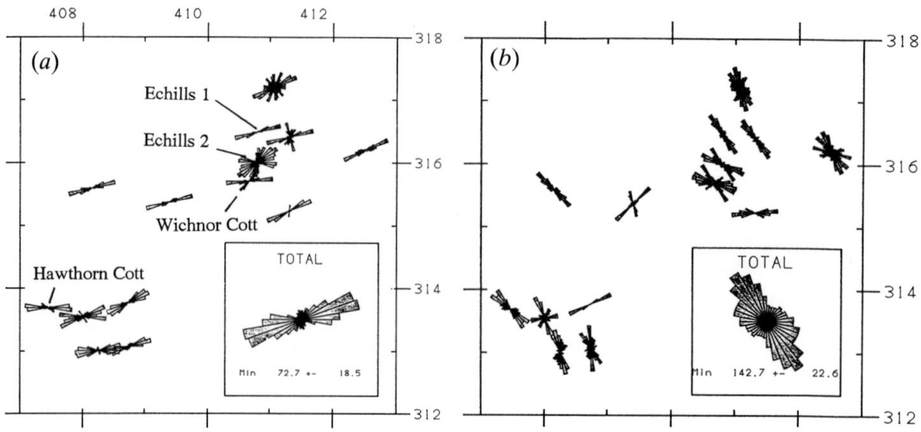

Figure 4. *Minimum* principal stress orientations (borehole breakout directions) for a study area in the English Midlands. Observations in (*a*) the Carboniferous are rotated by 70° compared with those in (*b*) the Permo-Triassic strata.

cluster of four boreholes in the NE corner of the study area. Maps of both the caliper and resistivity rose diagrams were very similar.

(*b*) Investigations of the Carboniferous and Permo-Triassic strata separately (figure 4) showed that the ENE trend is well defined in the Carboniferous strata (equivalent to an orientation of $\sigma_{\mathrm{H,max}}$ of $157 \pm 19°$), and the NNW trend is predominant in the Permo-Triassic strata. Further subdivisions of the Permo-Triassic into the Bunter Pebble Beds below and the Keuper Marl above results in a bimodal distribution for the Pebble Beds, with a generally NNW prevalence and a ENE trend for the Keuper Marl. Thus a *ca.* 70° rotation of stress orientation with depth, from NNW in the uppermost lithologies to ENE in the deeper strata, is evident.

(*c*) Low strength coal seams (FIDX values less than 30) were isolated in seven boreholes. The prevailing breakout orientation for these seams is the same as that for the Carboniferous as a whole. But individually, only in boreholes Echills 1 and 2 do all the seams show agreement with this trend. Nine seams in Hawthorn Cottage borehole show clear rotation with depth while in Wichnor Cottage borehole, the top three seams display a NE trend, reverting to ENE in the underlying four seams, below which eight seams show orientations of due east. These observed rotations are attributed to the relative proximity of particular seams to local faults.

Other interesting breakout behaviour was also observed, notably the following.

(*a*) It was common for the dipmeter tool to rotate as it passed through a coal seam (figure 5) and for the largest caliper eccentricities to occur at the contact between the coal and the underlying beds.

(*b*) Coal Measures breakouts showed a highly significant and consistent inverse correlation between rock strength and the magnitudes of both caliper and resistivity eccentricities: i.e. the weaker coal, seatearth and siltstone materials commonly produce the largest breakouts. However, this correlation was not so strong for the Permo-Triassic sediments.

(*c*) Towards the bottom of some boreholes there is a marked reduction in caliper eccentricities, relative to the corresponding resistivity eccentricities, demonstrating that in the upper borehole there are zones of coincident caliper and resistivity

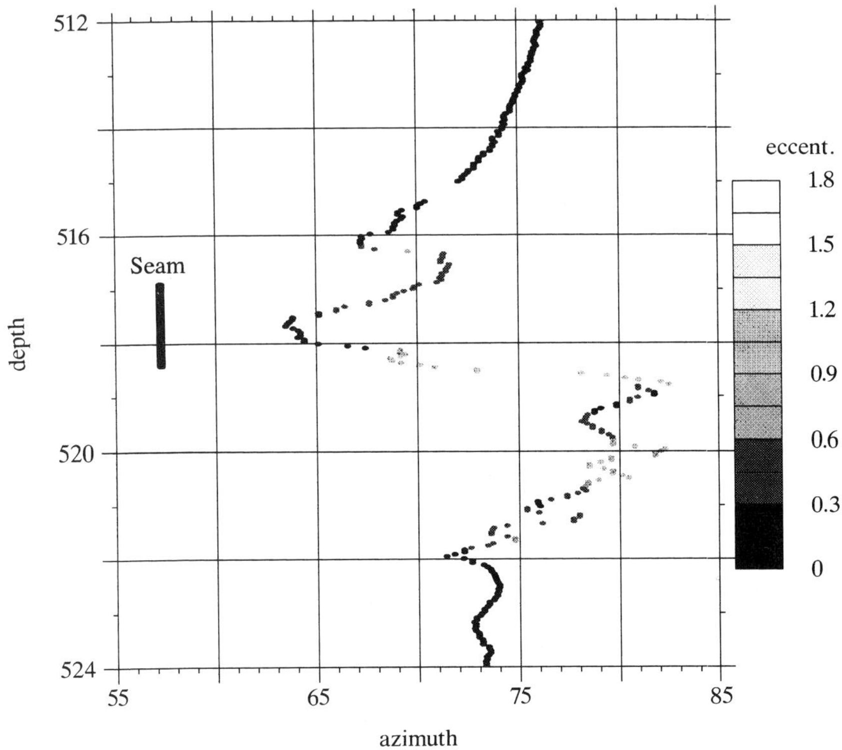

Figure 5. An example of stress rotation with depth (some 15°) across a coal seam. The figure shows the azimuth of breakout direction for the different eccentricities that are observed. The largest caliper eccentricity occurs at the contact between the coal and the underlying beds.

eccentricity where both microcrack propagation and material spalling have taken place, whereas in the lower sections the rock have not had time to dislodge. This observation (see also Kessels 1989) supports the suggestion that breakouts continue to develop with time and that borehole stability is not fully achieved during drilling.

6. Continuing work

It has been shown that Europe can be subdivided into three stress provinces within each of which the broad regional stress field is relatively consistent and indicative of major tectonic events. But to some extent this subdivision may be too simplistic because the data itself tends to be clustered into these same broad groupings. The diversity of stress orientations deduced, for example, from the Balkans indicates a very complex situation. In the Mediterranean region, where seismicity is closely related to plate-boundary effects, borehole breakout analysis is probably the main source of data for the study of the modern intraplate upper crustal state of stress. There are still big gaps in the database especially for much of eastern Europe, southern France, Spain, Italy and the Mediterranean. Even in the North Sea and other offshore areas around Britain, where the potential source of data is huge, the number of systematic borehole breakout analyses that have been performed is small.

Although regional-scale stress fields are relatively consistent, it has also been demonstrated that, when examined in detail, the stresses can be subject to the influences of small-scale local features and hence locally can be very variable. What is needed, therefore, is an expansion of the current database to provide not only a wide geographical distribution of stress data but also a high density distribution of data in three dimensions. The level of detail of the analysis of stress data will depend upon the application of the knowledge to be gained. The change of stress within and across a coal seam or around a salt dome is of considerable interest to the mining or hydrocarbons industry, as the case may be, and the time dependence of breakout evolution in boreholes may influence the economics of deep drilling. But it is the tectonic causes of, for example, stress rotations from one stratigraphic unit to another and the inter-relationships between stress and structure that are of primary interest to the research geologist.

Additional aspects of rock stress research which need to be taken further include (i) the influence of time-dependent breakout evolution on borehole stability; (ii) the use of borehole leak-off test data to help improve our knowledge of stress magnitude variations; (iii) the interdependence of rock stresses and interlayer pore pressure contrasts; their influence on the initiation and propagation of fractures and faults; and whether time-dependent pore pressure changes due to hydrocarbon production will cause changes in localized stress fields.

Collaborative studies are underway between the Universität Fridericiana Karlsruhe, the British Geological Survey and other institutes to help fill some of these database gaps and to improve our knowledge of stresses throughout Europe.

We acknowledge contributions from colleagues to the research reported here: Dr Chris Evans and Roger Peart (BGS, Keyworth); Mary Lou Zoback (USGS, Menlo Park); Holger Spann (Karlsruhe); and Keith Whitworth and Kevin Onions (British Coal Corporation). Thanks are also due to the Commission of the European Communities, BEB and the SFB 108 'Stress and Stress Release in the Lithosphere' for funding some aspects of this work and to British Coal for permission to use borehole data.

References

Ahorner, L. 1970 Seismotectonic relations between the graben zones of the Upper and Lower Rhine valley. In *Graben problems* (ed. H. Illies & St. Müller), pp. 155–166. Stuttgart: Schweizerbart.

Ahorner, L. 1975 Present-day stress field and seismotectonic block movements along major fault zones in western Europe. *Tectonophys.* **29**, 233–249.

Anderson, E. M. 1951 *The dynamics of faulting*. Edinburgh: Oliver and Boyd.

Bell, J. S. & Babcock, E. A. 1986 The stress regime of the western Canadian basin and implications for hydrocarbon production. *Bull. Can. Petroleum Geol.* **34**, 364–378.

Bell, J. S. 1989 The stress regime of the Scotian Shelf off-shore eastern Canada to 6 Kilometres depth and implications for rock mechanics and hydrocarbon migration. In *Proc. Rock at Great Depth, Pau* (ed. V. Maury & D. Fourmaintraux), pp. 1243–1265. Rotterdam: A. A. Balkema.

Bell, J. S. & Lloyd, P. F. 1989 Modelling of stress refraction in sediments around the Peace River Arch, western Canada. In *Current research*, Part D, Paper 89-1D, pp. 49–54. Geological Survey of Canada.

Bevan, T. G. & Hancock, P. L. 1986 A late Cenozoic regional mesofracture system in southern England and northern France. *J. geol. Soc. Lond.* **143**, 355–362.

Bradshaw, G. A. & Zoback, M. D. 1988 Listric normal faulting, stress refraction, and the state of stress in the Gulf Coast basin. *Geology* **16**, 271–274.

Brereton, N. R. & Evans, C. J. 1989 Rock stress orientations from borehole breakouts. In *European geothermal update. Proc. 4th Int. Seminar on the Results of EC Geothermal Energy Research and Demonstration, Florence* (ed. K. Louwrier, E. Staroste & J. D. Garnish), pp. 213–231. Kluwer Academic Publishers.

Brereton, N. R., Evans, C. J., Peart, R. J. & Hyett, A. J. 1990 Mechanisms and theoretical basis governing the creation and development of borehole breakouts. *Br. geol. Surv. Tech. Rep.* WK/90/18.

Bungum, H., Alsaker, A., Kvamme, L. B. & Hansen, R. A. 1991 Seismicity and seismotectonics of Norway and nearby continental shelf areas. *J. geophys. Res.* (In the press.)

Clauss, B. 1987 Numerische modellierung der tektonischen spannungen in der Nordschwiz. Diploma thesis, University of Karlsruhe.

Downing, R. A. & Gray, D. A. (eds) 1986 Geothermal energy – the potential in the United Kingdom. British Geological Survey Report.

Gale, W. L. & Blackwood, R. L. 1987 Stress distributions and rock failure around coal mine roadways. *Int. J. Rock Mech. Min. Sci. Geomech. Abstr.* **24**, 165–173.

Goodier, J. N. 1933 Concentration of stress around spherical and cylindrical inclusions and flaws. *Trans. Soc. Am. mech. Engng* **55**, 39–47.

Halker, A., Kuznir, N. J., Mellor, D. W. & Whitworth, K. R. 1982 The synthesis of fracture/strength logs using borehole geophysics: a new geotechnical service. *Q. Jl Engng Geol. Lond.* **15**, 15–28.

Hyett, A. J. 1990 The potential state of stress in a naturally fractured rock mass. Ph.D. thesis, University of London, Imperial College of Science, Technology and Medicine.

Ivins, E. R. & Lyzenga, G. A. 1986 Stress patterns in an intraplate shear zone: an effective anisotropic model and implications for the transverse ranges, California. *Phil. Trans. R. Soc. Lond.* A **318**, 285–347.

Jackson, P. J., Fairbairn, C. M. & Whitworth, K. R. 1989 Geophysical exploration in the British coalfields. *Min. Res. Engng* **2**, 123–149.

Liu, X. & Zoback, M. D. 1991 The effect of topography on the state of stress at depth in the crust *J. geophys. Res.* Special issue on the Cajon Pass Scientific Drilling Project. (In the press.)

Kessels, W. 1989 Observation and interpretation of time-dependent behaviour of borehole stability in the Continental Deep Drilling pilot borehole. *Scientific Drilling* **1**, 127–134.

Kusznir, N. J. & Bott, M. H. P. 1977 Stress concentration in the upper lithosphere caused by underlying viscoelastic creep. *Tectonophys.* **42**, 247–256.

Minster, J. B. & Jordan, T. H. 1978 Present-day plate motions. *J. geophys. Res.* **83**, 5331–5354.

Müller, B., Fuchs, K., Mastin, L., Gregersen, S., Pavoni, N., Sephansson, O., Ljunggren, C. & Zoback, M. L. 1991 Regional patterns of stress in Europe. *J. geophys. Res.* (In the press.)

Shamir, G. & Zoback, M. D. 1989 The stress orientation profile in the Cajon Pass, California, scientific drillhole, based on detailed analysis of stress induced borehole breakouts. In *Proc. Rock at Great Depth, Pau* (ed. V. Maury & D. Fourmaintraux), pp. 1041–1048. Rotterdam: A. A. Balkema.

Schneider, T. 1985 Bohrlochrandausbrueche in norddeutschen bohrungen und ihm bezichung zum regioalen spannungensfeld – beobachtung und theorie, Diploma thesis, University of Karlsruhe.

Springer, J. E. 1987 Stress orientations from wellbore breakouts in the Coalinga region. *Tectonics* **6**, 667–676.

Stephansson, O. 1988 Ridge push and glacial rebound as rock stress generators in Fennoscandia. *Bull. geol. Inst. Univ. Uppsala, N.S.* **14**, 39–48.

Stephansson, O., Särkkä, P. & Myrvang, A. 1986 State of stress in Fennoscandia. In *Proc. Int. Symp. on Rock Stress and Rock Stress Measurements* (ed. O. Stephansson), pp. 21–32. Stockholm: Centek.

Stein, S., Cloetingh, S., Sleep, N. & Wortel, R. 1989 Passive margin earthquakes, stresses and rheology. In *Earthquakes at North-Atlantic passive margins – neotectonics and postglacial rebound* (ed. S. Gregersen & P. W. Basham), pp. 231–259. NATO ASI Series C. Kluwer.

Swolfs, H. S. & Savage, W. Z. 1985 Topography, stresses and stability at Yucca Mountain, Nevada. *26th US Symposium on Rock Mechanics, Rapid City*, pp. 1121–1129.

Udias, A., Buforn, E. & de Gauna, R. 1989 Catalogue of focal mechanisms of European earthquakes. Report of the Department of Geophysics, Universidad Complutense, Madrid.

Zoback, M. L. *et al.* 1989 Global patterns of tectonic stress. *Nature, Lond.* **341**, 291–298.

Discussion

P. Hancock (*Bristol University, U.K.*). I would like to report some unpublished overcoring measurements. The first set were made in the central valley of Scotland and the second set were made in the middle of the Orcadian Plateau in Caithness. They gave reliable results showing a consistent NW–SE trend. In the Permo-Triassic rocks of the Midlands you showed one set of breakout directions in the Bunter Pebble Beds and another in the Keuper rocks. I wonder whether there is much greater dispersion of the data within the Bunter Pebble Beds because they are so extremely heterogeneous?

R. Brereton. The two breakout trends from the Midlands were from the Carboniferous and Permo-Triassic strata. Within the Permo-Triassic rocks there is a breakout trend which rotates by about 70° from NNW to ENE with depth. However, the dispersion of data within the Bunter Pebble Beds (Sherwood Sandstone Group) was not markedly different from that for the Keuper rocks (Mercia Mudstone Group).

T. Harper (*BP Research Centre, Sunbury, U.K.*). Are there any magnitude data from the coalfield where there are a lot of closely spaced directional data which might enable a generic model for that distribution to be developed?

R. Brereton. There is as yet insufficient stress magnitude data to enable a model to be developed at this stage.

M. H. P. Bott (*Durham University, U.K.*). With respect to possible reorientations of the stress system about a salt dome which Dr Brereton discussed, might this not also apply in the Alps, the Alps being a weak surface in the lithosphere?

B. Müller. If the Western Alps, which show a roughly radial pattern of maximum horizontal stress with respect to the Alpine front, acts as a soft inclusion in the lithosphere we would expect to find (1) relatively homogeneous stress orientations within the Alps and (2) stress orientations which trend more or less parallel to the Alpine front outside the Alps. The available data do not show such a tendency. N. Pavoni interpreted the radial pattern as being due to a large-scale crustal indentation of the inner parts of the Alps towards the foreland in the west. Alternatively we suggest that the dense and thick lithosphere root could overcome the tension caused by the high topography and create an additional compression perpendicular to the trend of the lithosphere root.

D. Sanderson (*Southampton University, U.K.*). I am interested in the breakout results which showed a difference between the Trias and the Carboniferous because the two directions which Dr Brereton showed, the NW or NNW direction in the Trias and the more ENE direction in the Carboniferous, are precisely parallel to the main directions of extension fractures which developed in the Carboniferous and the Trias. I wonder whether the breakouts were not utilizing some pre-existing anisotropy.

R. Brereton. There has been some debate in the literature as to whether breakouts reflect pre-existing fracture patterns or not. To some degree this question is about whether the stresses being measured today are modern or residual and if they are residual whether the observed fracture patterns in the rocks are themselves some manifestation of the residual stress field. If so some correspondence between the two sets of results might be expected. In an earlier paper Professor Kusznir discussed a stress 'memory effect' on a continental scale implying that there may indeed be a strong residual component to the stress orientations being measured by breakouts. In my view this debate has not yet been fully resolved.

N. Kusznir (*Liverpool University, U.K.*). I find the difference in orientation of stress between Scandinavia and the rest of NW Europe fascinating. Could Dr Müller explain this change in orientation?

B. Müller. A combination of different sources seems to be responsible for the variability of the stress orientations in Scandinavia: (*a*) ridge-push forces from the mid-Atlantic Ridge, possibly also an additional radial component of the Iceland Hot Spot; (*b*) flexure due to the deglaciation of wide areas of Scandinavia which may cause a tensile component in the area of deglaciation and a compressional component in the surrounding oceanic crust, both oriented radial to the area of deglaciation; (*c*) topography might be responsible for local variations. The magnitude of the broad-scale tectonic forces, e.g. the ridge-push forces, is strongly influenced by the variation of lithospheric thickness, causing stress amplification in those parts of the lithospheric plates with thin lithosphere. Areas of thick lithosphere show lower stress magnitudes that might be much more easily disturbed.

Stress magnitudes in the crust: constraints from stress orientation and relative magnitude data

By Mary Lou Zoback[1]† and Marian Magee[2]

[1] *Geophysikalisches Institut, Universität Karlsruhe, Hertzstrasse 16, 7500 Karlsruhe 21, F.R.G.*

[2] *U.S. Geological Survey, MS 977, 345 Middlefield Road, Menlo Park, California 94025, U.S.A.; Department of Geophysics, Stanford University, Stanford, California 94305, U.S.A.*

The World Stress Map Project is a global cooperative effort to compile and interpret data on the orientation and relative magnitudes of the contemporary *in situ* tectonic stress field in the Earth's lithosphere. Horizontal stress orientations show regionally uniform patterns throughout many continental intraplate regions. These regional intraplate stress fields are consistent over regions 1000–5000 km wide or *ca.* 100 times the thickness of the upper brittle part of the lithosphere (*ca.* 20 km) and about 10–15 times the thickness of typical continental lithosphere (*ca.* 150–200 km).

Relative stress magnitudes or stress régimes in the lithosphere are inferred from direct *in situ* stress measurements and from the style of active faulting. The intraplate stress field in both the oceans and continents is largely compressional with one or both of the horizontal stresses greater than the vertical stress. The regionally uniform horizontal intraplate stress orientations are generally consistent with either relative or absolute plate motions indicating that plate-boundary forces dominate the stress distribution within the plates. Since most regions of normal faulting occur in areas of high elevation, the extensional stress régimes in these areas can be attributed to superimposed bouyancy forces related to crustal thickening and/or lithosphere thinning; stresses derived from these bouyancy forces locally exceed mid-plate compressional stresses.

Evaluating the effect of viscous drag forces acting on the plates is difficult. Simple driving or resisting drag models (with shear tractions acting parallel or antiparallel to plate motion) are consistent with stress orientation data; however, the large lateral stress gradients across broad plates required to balance these tractions are not observed in the relative stress magnitude data. Current models of stresses due to whole mantle flow inferred from seismic tomography models (and with the inclusion of the effect of high density slabs) predict a general compressional stress state within continents but do not match the broad-scale horizontal stress orientations. The broad regionally uniform intraplate stress orientations are best correlated with compressional plate-boundary forces and the geometry of the plate boundaries.

1. Introduction

A global compilation of *in situ* stress data is currently underway as part of the World Stress Map (WSM) project, a Task Group of the International Lithosphere Program.

† Permanent address: U.S. Geological Survey, MS 977, 345 Middlefield Road, Menlo Park, California 94025, U.S.A.

The goal of this project is to compile and interpret data on the intraplate tectonic stress field. More than 5700 data points on stress orientation and relative magnitudes (and absolute stress magnitudes where available) in the upper crust have been compiled to date. The focus of the WSM effort is to define the stress field within the plates and within broad deformational zones adjacent to plate boundaries; data on the stress state along plate boundaries (e.g. mid-ocean ridges or along or within Benioff zones) have not in general been compiled as the kinematics of these boundaries are generally well known. Through interpretation of the intraplate stress data, constraints can be placed on the magnitudes of both broad scale and local sources of stress acting on the lithosphere.

Results to date of the WSM project will be summarized in a special issue of *J. geophys. Res.*, now nearly completed. The special issue contains over 20 papers covering all aspects of the project from description of the data, discussion of regional stress patterns and their relationship to tectonics, as well as finite element modelling of the stress data to constrain the magnitudes of forces acting on the plates.

The purpose of this paper is to investigate a specific aspect of stress in the lithosphere, the implications of the global stress database for the magnitude of stress in the upper lithophere. Relative stress magnitudes are inferred from the style of faulting observed in over 3032 intraplate focal mechanisms compiled worldwide, allowing a characterization of mean stress levels within the plates relative to the lithostat.

Following early finite element modelling of global intraplate stress patterns by Richardson *et al.* (1979), numerous modelling papers have demonstrated that most of the broad-scale stress patterns observed within the interiors or plates can be attributed directly to the effects of plate boundary forces (see summary of recent modelling attempts in Zoback (1991*a*)). Drag forces typically are assumed to be resistive and of small magnitude and used to balance the torque on individual plates (Richardson *et al.* 1979; Cloetingh & Wortel 1986; Richardson & Reding 1991). Recently Bai *et al.* (1991) have calculated the stress effect in the lithosphere of drag related to flow induced by deep mass inhomogeneities in the mantle inferred from global seismic tomography models. The possible influence on the intraplate stress field of this drag, which is related to large-scale mantle flow, is evaluated by a comparison between the WSM observations and the stress patterns (both orientation and relative magnitude) predicted by Bai *et al.* (1991).

2. First-order patterns of stress in the lithosphere

The available global stress orientation data are shown in figure 1, which plots the maximum horizontal stress ($S_{H,max}$) orientation. In areas of extensional stress $S_{H,max}$ corresponds to the intermediate stress S_2; in all other regions $S_{H,max}$ corresponds to the maximum stress, S_1. The data are derived from a variety of geologic and geophysical indicators: the P- and T-axes of earthquake focal mechanisms, well-bore breakouts, *in situ* stress measurements by hydraulic fracturing and overcoring, volcanic vent alignments, and fault slip data. The geologic data included are generally of Quaternary age (or in regions of possible multiple Quaternary deformation episodes, the data correspond to the youngest deformation). In some stable areas (such as the eastern United States) fault slip data as old as post-Miocene have been included. Note that the lengths of the oriented lines shown on figure 1 correspond to quality or reliability of the data as a tectonic stress indicator. This

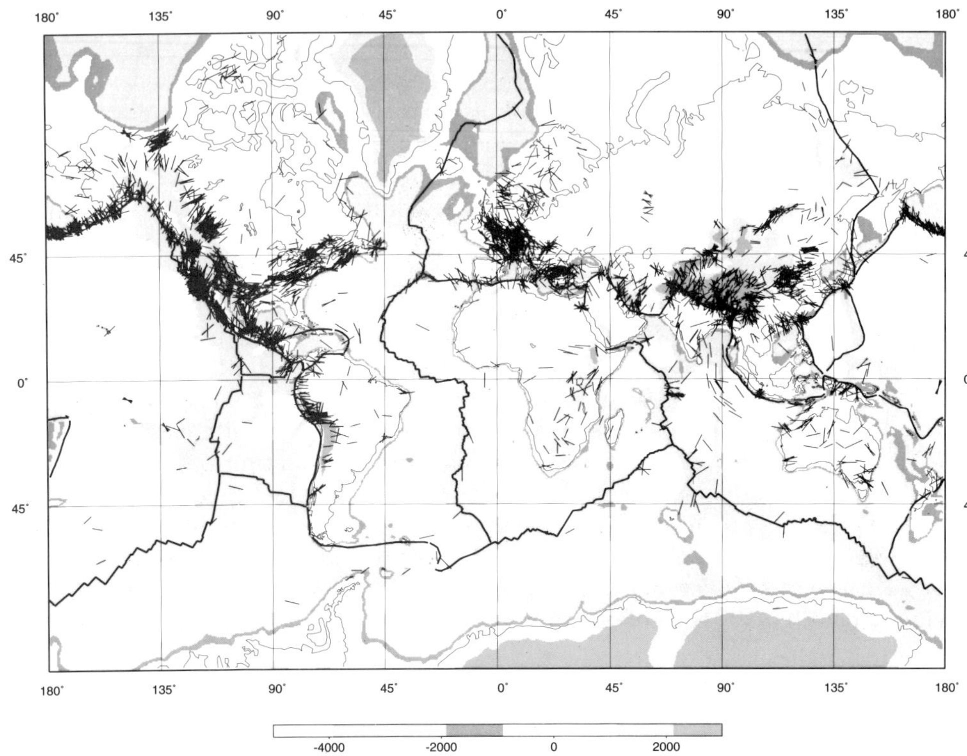

Figure 1. World Stress Map, Mercator projection. $S_{H,max}$ orientations are plotted, length of lines are proportional to quality.

quality ranking system is based on accuracy of the data, the depth interval sampled and the general reliability of the technique used. A detailed description of the theory and limitations of these various stress indicators and the quality ranking system has been given in several recent papers (Zoback & Zoback 1989, 1991; Zoback 1991*a*), the reader is referred to these papers for additional information. A large size colour map of the data shown here in figure 1 is included in Zoback (1991*a*).

Figure 2 is a generalized stress map derived from mean stress orientations for regions of dense data coverage. Absolute velocity trajectories for individual plates computed using absolute velocity poles AM1-2 of Minster & Jordan (1978) are also given. Many of the general conclusions of the global data compilation based on stress orientations (Zoback *et al.* 1989; Zoback 1991*a*) can be verified by inspection of figures 1 and 2.

1. Stress orientations inferred from different techniques yield consistent and similar orientations despite the different depths and volumes of rock sampled, the data indicate that both the orientation and relative magnitude of the stress field are generally uniform in depth through brittle layer.

2. The different stress indicators define a tectonic stress field and indicate that broad regions of the Earth's crust (1000s km on a side) are subject to uniform patterns of stress orientation, despite geologic and structural complexity of the upper crust, the source of these regionally uniform stress fields must be broad scale.

3. There is a correlation between $S_{H,max}$ orientations and the directions of absolute plate motion for broad regions of many plates.

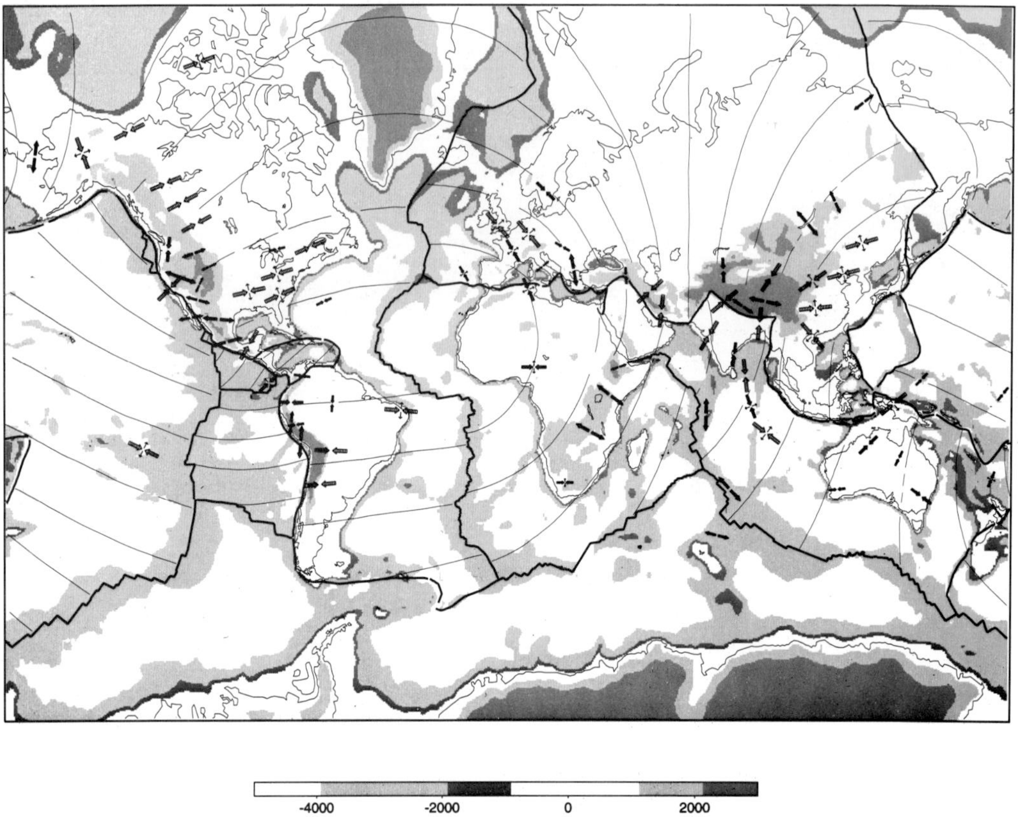

Figure 2. Generalized stress map of the world. Large arrows indicate mean stress directions for dense clusters of data. Inward pointing arrows indicate $S_{H,max}$ orientations for compressive stress regimes where $S_{H,max}$ coincides with S_1 and one or both of the horizontal stresses exceeds the vertical stress. Outward pointing arrows show $S_{H,min}$ orientations. Thick outward pointing arrows indicate regions of extensional stress régime where both horizontal stresses are less than the vertical stress. Thin outward pointing arrows plotted together with inward pointing arrows indicate a strike–slip faulting stress régime. Coverage is non-uniform due to the non-uniform data coverage in figure 1. Light lines are absolute velocity trajectories from model AM1–2 of Minster & Jordan (1978).

Finite-element modelling to predict intraplate stresses from both body forces and tractions acting on the plates has now been completed for a number of individual plates. The consistent ENE $S_{H,max}$ direction in the eastern United States and throughout much of Canada is generally parallel to the direction of absolute motion of North America (Zoback *et al.* 1986); however, Richardson & Reding (1991) demonstrate that this direction is also consistent with ridge push forces. Similarly, $S_{H,max}$ orientations in South America agree both with ridge push and absolution motion directions (Assumpcao 1991). Meijer & Wortel (1991) and Jurdy & Stefanick (1991) both conclude that ridge push is the dominant force responsible for the E–W compression in South America, with local influence of trench suction or buoyancy forces causing extension in the Andean region. Meijer & Wortel (1991) also concluded that driving drag could not be ruled-out as an important force affecting South America; they did not, however, compute the predicted stresses for a driving drag model.

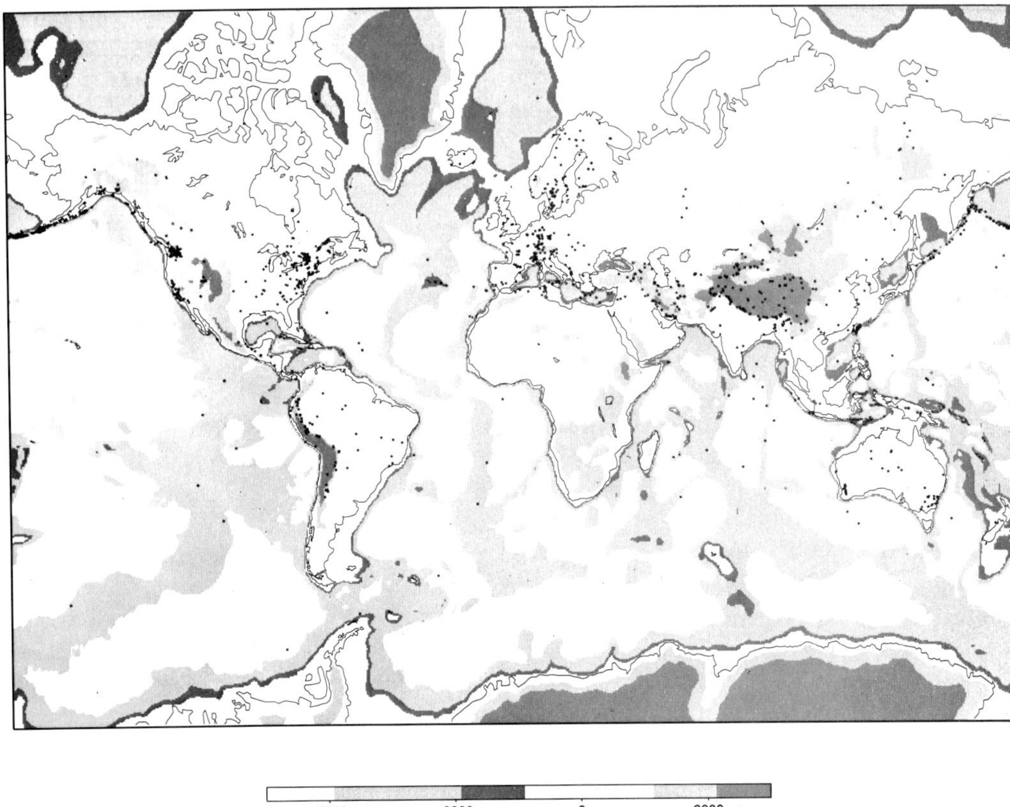

Figure 3. Location (by small crosses) of all data indicating a thrust faulting stress régime ($S_{H,max} > S_{H,min} > S_v$) in the WSM database.

$S_{H,max}$ orientations are subparallel to absolute plate motion in western Europe where the absolute plate velocity is poorly resolved; however, an even better correlation is observed between $S_{H,max}$ orientations and the relative velocity trajectories for the convergence of Africa and Eurasia (Müller *et al.* 1991). The NNW to NW compression throughout western Europe (excluding the Mediterranean region) has been successfully modelled as the combined effect of ridge push along the northwestern border of the plate and convergence and continental convergence and collision between Africa and Eurasia along the southern border (Grunthal & Stromeyer 1991).

Sparse stress data from western and southernmost Africa indicate E–W compression consistent with ridge-push directions and nearly orthogonal to the absolute motion direction (Zoback 1991*a*). However, in eastern Africa, the intraplate stress field is dominated by extensional stresses related to buoyancy forces derived from a thin lithosphere and warm upwelling upper mantle (see §3*a*).

The stress pattern in Australia is complex; however, the best data in central Australia suggest NNE to NE compression approximately in the direction of absolute motion, but also consistent with ridge push along the southern boundary of the plate and arc collision along the northern boundary. E–W compression is well documented in both southeastern and southwestern Australia and is difficult to explain in terms of plate boundary forces.

In other places, $S_{H,max}$ directions do not correlate directly with absolute or relative plate motion but are clearly related to plate-boundary forces. For example, the radial pattern of $S_{H,max}$ orientations in eastern Eurasia fanning out from the Himalayan collision have been successfully modelled by a number of workers as the result of collisional forces related to the indentor geometry of India into Eurasia (a recent example of such modelling is given in England & Houseman (1989)). Thus once again the influence of forces acting along the plate boundary (in this case related to the Eindentor of India into Eurasia (Molnar & Tapponier 1975)) is seen to be the dominant force influencing the stress field 1000s of km from the plate boundary.

Clearly plate driving forces play a critical role in determining the stress field in the interior of the plates. However, the geometry of the plate boundary apparently plays an equally important role in determining the actual orientation of the intraplate stress field. Two compressive forces, continental collision and ridge push (particularly for plates with no or only a very small attached slab), play a dominant role in determining the intraplate stress field as first noted by Richardson *et al.* (1979). Net slab forces (balance of slab pull and both viscous and frictional resistance to subduction) are probably generally small and compressive (Richardson *et al.* 1979; Cloetingh & Wortel 1986) as demonstrated by an absence of extensional intraplate earthquakes in old oceanic crust (Wiens & Stein 1985). The role of drag forces is more difficult to evaluate since, as Richardson (1991) points out, the ridge push and absolute motion torque poles are nearly identical for most plates; thus effects of the ridge push and either driving drag or resisting drag in the direction of absolute plate motion can not be differentiated on the basis of stress orientations alone. However, there are considerable differences in the predicted stress magnitudes for drag versus ridge-push forces. Stress magnitudes due to ridge push are compressive constant throughout the continents and in old oceans (where oceanic topography no longer follows the t^{-2} cooling curve); however, tractions across the base of the plate due to drag induce stress gradients across the plates. Thus to properly evaluate drag we need information on stress magnitudes.

3. Magnitude of stress in the lithosphere

The World Stress Map project has systematically compiled data on stress orientations; however, since the stress field is a tensor, complete characterization requires magnitude information. Available stress magnitude determinations (from *in situ* stress measurements) have also been compiled; however, these data represent only about 2% (103/5701) of the database (and generally sample only the upper 1–2 km of the Earth's crust). Because both technical and economic constraints limit the depth of stress magnitude measurements to generally less than seismogenic depths (5–20 km), we must infer absolute stress magnitudes either from laboratory studies, or from downward extrapolation of near-surface stress measurements.

However, information on relative or absolute stress magnitudes obtained from earthquake focal mechanisms, *in situ* stress measurements, and young fault slip data can be used to categorize the data into three main stress régimes: normal faulting, strike–slip faulting, and thrust faulting (as defined below). In figures 3–5 the location of data corresponding to each of these three régimes is indicated by small crosses.

For data where information on the plunges of either the *P*-, *B*- and *T*-axes or S_1, S_2, and S_3 axes are available, specific ranges of plunges were used to categorize the data into the three stress régimes. The rationale for these specific ranges of plunges

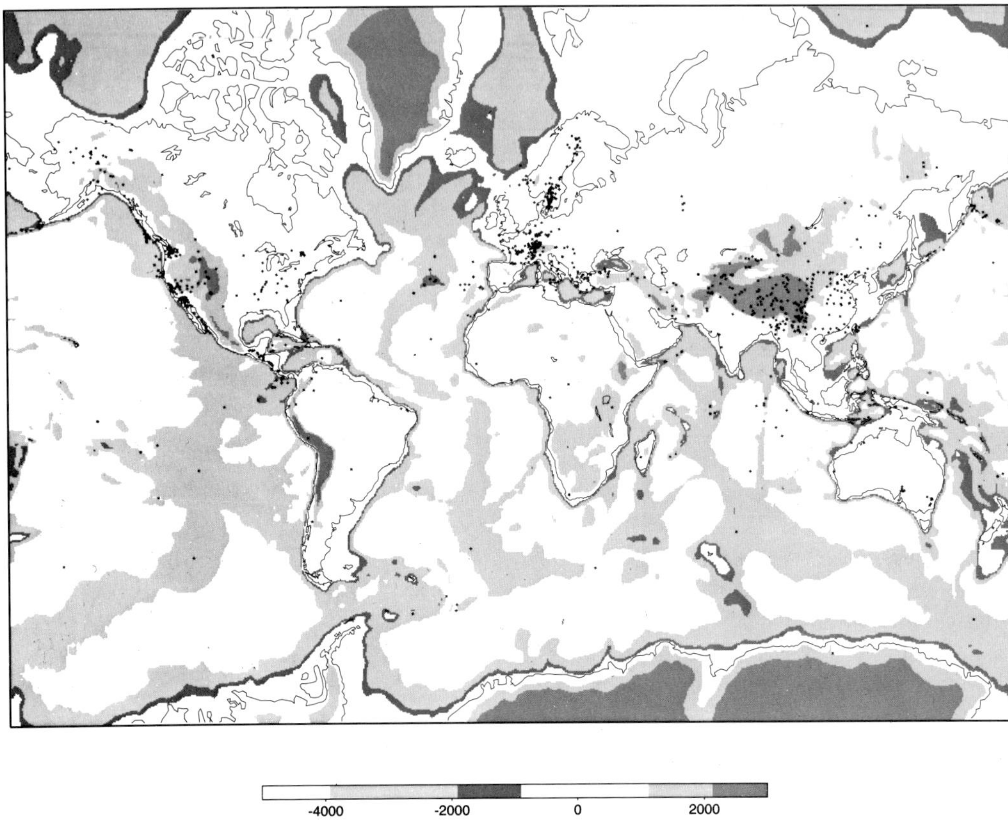

Figure 4. Location (by small crosses) of all data indicating a strike–slip faulting stress régime ($S_{H,max} > S_V > S_{H,min}$) in the WSM database.

is based on an assumption of principal stresses in the crust lying in approximately horizontal and vertical planes, the actual values somewhat subjective (see discussion in Zoback 1991 *a*). The three main stress régimes and style of faulting (and associate ranges of axes' plunges) are:

	P/S_1 plunge	B/S_2 plunge	T/S_3 plunge
(1) extensional stress régime ($S_V > S_{H,max} > S_{H,min}$), normal dip slip (figure 5)			
	$pl > 52°$,		$pl < 35°$,
or	$40° < pl < 52°$,		$pl < 20°$;
(2) strike–slip stress régime ($S_{H,max} > S_V > S_{H,min}$), dominantly horizontal slip (figure 4)			
	$pl < 40°$,	$pl > 45°$,	$pl < 20°$,
or	$pl < 20°$,	$pl > 45°$,	$pl < 40°$,
(3) thrust stress régime ($S_{H,max} > S_{H,min} > S_V$), reverse dip slip (figure 3)			
	$pl < 20°$,		$40° < pl < 52°$,
or	$pl < 35°$,		$pl > 52°$.

Another general conclusion of the WSM project is clear when comparing figures 3–5: most midplate regions are characterized by compressional deformation (strike–slip and thrust faulting stress régimes), regions of extensional stress régime occur primarily in topographically high areas in both the continents and the oceans.

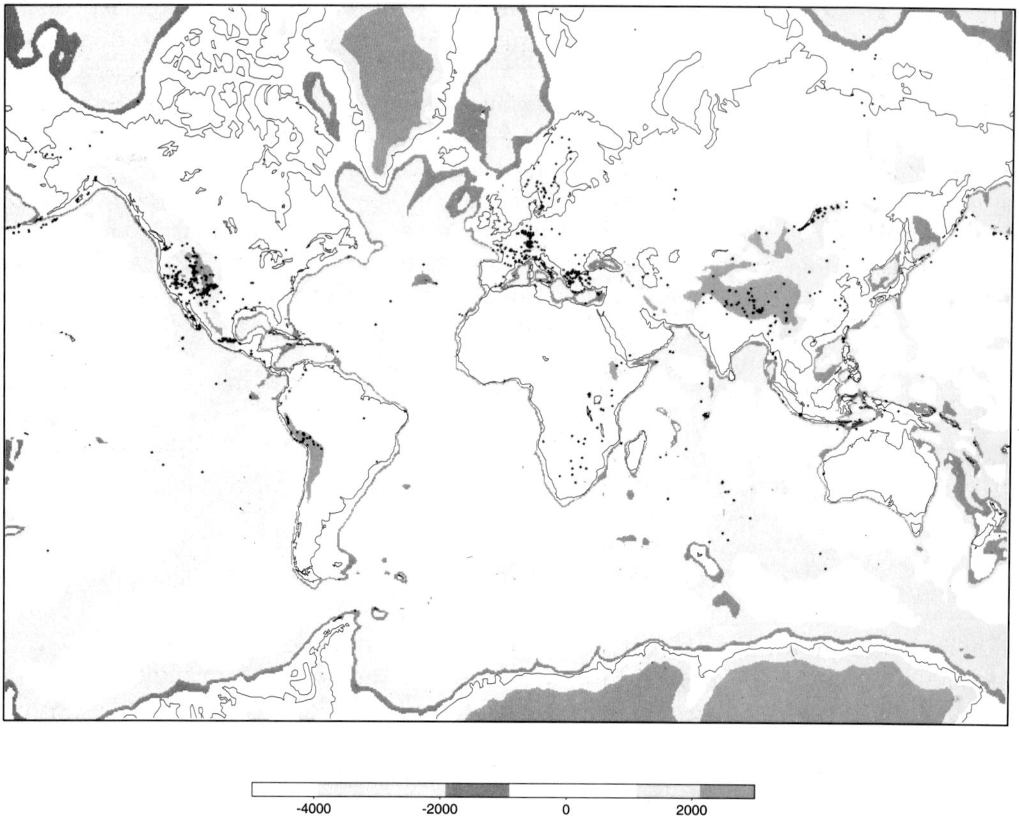

Figure 5. Location (by small crosses) of all data indicating a normal faulting stress régime ($S_V > S_{H,max} > S_{H,min}$) in the WSM database.

In detail, a number of conclusions can be drawn about relative stress magnitudes in the lithosphere by comparing the distribution of data shown on figures 3–5. As can be seen in figure 3, much of the Australian continent is dominated by thrust régime. Similarly thrust faulting dominates in southeastern Canada (possibly a local effect related in part to glacial rebound influence (Zoback 1991*b*)), throughout continental South America, and in a broad zone of continental convergence in the Caucasus. Dominantly strike–slip deformation (figure 4) is seen along the San Andreas system, western Africa, and eastern China, and in continental Alaska. Comparison of figures 3 and 4 indicate that many other midplate areas are subject to a combination of strike–slip and thrust faulting régimes: much of the central and eastern United States, the Pacific Northwest of the United States (region overlying the subducted Juan de Fuca plate), old parts of the Pacific plate, the eastern Indian ocean and India subcontinent, Fennoscandia, western Europe, and the Tibetan Plateau. As discussed below, these last two areas also contain a fair amount of normal faulting data but as can be seen in figure 1 the $S_{h,max}$ orientations are constant for data in all three régimes.

The normal faulting stress data mostly occur in elevated regions: western U.S. Cordillera, the High Andes, East African rift, Tibetan plateau, Baikal rift, and the western Indian Ocean. Other areas of extension occur in topographically low regions, most notably the Aegean region which lies in a back-arc setting. There are also a large number of normal faulting earthquakes in western Europe, well north of the Alps. This is in marked contrast to midplate North America where normal faulting is quite rare.

In some cases the source of the high elevation in some of the extensional regions may be simply related to thickened crust (Andes and the Tibetan Plateau); however, in both these regions the elevation has been attributed to combined effects of a thickened crust and a thinned mantle lithosphere (see Isacks (1988) for discussion of upper mantle structure beneath the Andean region and England & Houseman (1989) for a discussion of the upper mantle contribution to explain both rapid uplift and the extensional state of stress on the Tibetan plateau). In many other regions (such as the western U.S., East Africa, and Baikal rift, the crust is quite thin (30–35 km thick) and the source of the high elevation is clearly related to a thinned mantle lithosphere and upwelling hot asthenosphere (Thompson & Zoback (1979) for western U.S., Henry *et al.* (1990) for the East African rift, and Zorin & Rogozhina (1978) for the Baikal rift). Thus the state of stress in many extensional regions can be related to the local effects of bouyancy forces derived not only from a crustal root, but in many cases dominated by the effects of thermal upwelling and thinned lithosphere (as described below).

(*a*) *Role of buoyancy forces*

It is clear from the general pattern of midplate compression that areas of extension require additional superimposed forces to produce the normal faulting stress régime in which both horizontal stresses are least than the vertical stress. In fact as can be seen in figure 5, the direction of extension in many of the extensional regions is generally perpendicular to the trend of the topographic bulge indicating the importance of the buoyancy force in overcoming the regional midplate compressional stress state. This observation can be used to place limits on the average magnitude of the mid-plate compression in the uppermost lithosphere due to plate driving forces such as ridge push and continental collision (Artyushkov 1973; Fleitout & Froidevaux 1982, 1983).

The horizontal compression of the ridge-push force is a result of the horizontal density contrast associated with the cooling and thickening of the oceanic lithosphere with age. Because the ocean floor topography and densities are generally well known, the ridge-push force is the easiest force to quantify and estimates of the total force range from $2\text{–}3 \times 10^{12}$ N m^{-1} (Frank 1972; Lister 1975; Parsons & Richter 1980), equivalent to *ca.* 20–30 MPa of stress averaged over a 100 km thick lithosphere. The actual magnitude of stress induced in the uppermost lithosphere depends upon how this force is distributed; the ductile nature of the lower part of the lithosphere causes a redistribution of any stress applied to the whole lithosphere and results in a stress amplification in the upper, more brittle part of the lithosphere (Kusznir & Bott 1977). For realistic rheologies, Kusznir & Park (1984, 1987) estimate a stress amplification by a factor of about 2–3 in the uppermost continental lithosphere for an applied compressional stress with surface heat flow of 60 mW m^{-2}.

Estimating stresses related to lateral density contrasts in regions of elevated topography is critically dependent on the assumptions about the density distribution within the lithosphere (England & Houseman 1989). However, lateral variations in this bouyancy-related stress are directly related to variations in geoid height; Froidevaux & Isacks (1984) use the geoid anomaly associated with the 4000 m high Altiplano-Puna plateau of the Andes to compute a bouyancy related force of between

4 and 5×10^{12} N m^{-1}. This force is thus comparable with, and possibly somewhat larger than, the ridge-push force for the large continental plateaus. Again, the actual stress magnitude in the upper lithosphere induced by this bouyancy force depends on lithosphere rheology and heat flow. Kusznir & Park (1987) compute stress amplification factors of as much as 8–11 for the middle crust for regions with surface heat flows of 95 mW m^{-2}.

Taking the stress amplification factor into account, it thus appears likely that bouyancy-related stresses may locally dominate the state of stress within the plate. Observed stress rotations (relative to regional, mid-plate orientations) in the vicinity of elevated regions can be used to place limits on the ratio of the magnitudes of regional versus local (bouyancy) stresses in the crust. Zoback (1991*a*) calculated that local bouyancy-derived extensional stresses in the upper lithosphere of the East African rift region must be roughly four times the magnitude of the regional stress differences in the plate to cause the roughly 60° rotation of the regional intraplate strike–slip stress régime with E–W $S_{H,max}$ to the observed normal faulting stress régime with a NNE trending $S_{H,max}$ in the vicinity of the East African rift. In many areas the superimposed effect of this bouyancy force may be more subtle. In the Colorado Plateau of the western United States, negative bouyancy forces associated with a relatively dense keel of upper mantle lithosphere have been suggested as the source of a 90° rotation of horizontal stresses between the extensional stress régimes of the Colorado Plateau and the surrounding Basin and Range (Zoback & Zoback, 1989).

As mentioned above, there are many normal faulting earthquakes in western Europe located distant from the Alpine belt and unrelated to any topographic effects due to that mountain chain. The stress orientation data in this region suggest that the origin of the mid-plate stress field in western Europe is largely related to compressional forces (ridge push and continental collision), thus dominantly strike–slip and thrust faulting earthquakes would be expected, as seen in mid-plate North America. Müller *et al.* (1991) conclude that local buoyancy forces related to mantle upwelling (e.g. beneath the Upper Rhine graben embayment and the Massif Central) may be responsible for the altered stress state.

(*b*) *Influence of drag forces*

As the lithospheric plates form the lid of a complex system of convection within the Earth's interior, forces resulting from relative motion the lithospheric plates and the upper mantle is another potential source of stress within the plates. There are several possible models of drag to choose from: resistive drag at the base of the plate opposing plate motions, so-called 'driving' drag in the direction of plate motion, drag due to 'counterflow' (mass flow from trenches back to ridges) or, drag due to mantle flow arising from density inhomogeneities throughout the mantle. As mentioned previously, resistive drag has generally been modelled as small and largely used to balance torque on the plates, the predicted $S_{H,max}$ orientations are antiparallel to the direction of absolute plate motion. The magnitude of the resistive drag forces are generally assumed proportional to the product of velocity, area, and a drag coefficient which may differ between cold thick continental lithosphere and thin, relatively warm oceanic lithosphere. Resistive drag would thus be expected to have greatest effect on fast moving, old continental plates such as Australia; however, evidence in central Australia for a correlation between absolute plate motion and $S_{H,max}$ orientation is weak (and could alternately be explained as a result

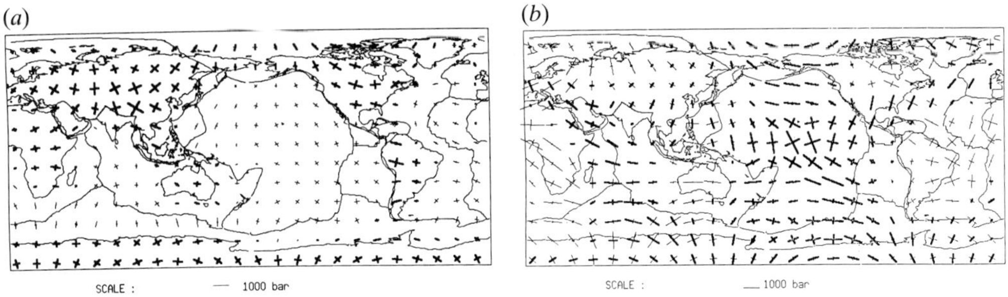

Figure 6. (*a*) Deviatoric stress patterns predicted from variations between oceanic and continental lithospheric density structure (expanded up to degree $l = 15$). The thick lines refer to extension, the thin ones to compression. (*b*) Superposition of 'reference' stress state (shown in (*a*)) and stresses arising from basal shear coupling due to mantle flow derived from a model driven by mass heterogeneities defined by seismic tomography and slab distribution. Figure from Bai *et al.* (1991) used with permission. (1000 bar = 10^8 Pa).

of ridge push and arc collision as explained above (see also Richardson 1991). Furthermore, zones of E–W compression on southeastern and southwestern coast of Australia would seem to rule this out.

Effects of counterflow drag are more difficult to evaluate. Counterflow calculations based on return of mass from the Aleutians to the mid-Atlantic ridge predict NW–SE directed flow beneath eastern North America (Chase 1979; Hager & O'Connell 1979). This is in clear disagreement with the well-constrained broad region of ENE trending $S_{H,max}$ directions in this region and implies that the counterflow must occur at a level too deep to influence stress states in the lithosphere (Zoback *et al.* 1986).

Models incorporating simple driving drag yield principal stress directions parallel and perpendicular to absolute plate motion and predict a stress gradient across the plate. Stress magnitudes due to drag are zero at the ridges and increase across the plate in the direction of absolute motion (Richardson *et al.* 1979; Richardson & Reding 1991). Inspection of the relative stress magnitude data shown in figures 3–5 show no clear evidence for systematic changes in the stress régime across plates except in North America, where normal faulting in the western United States is clearly a superimposed effect related to bouyancy forces derived from the warm upper mantle. Systematic drag-related stress gradients would not be expected for the Eurasian plate since its motion is so slow (barely resolvable in a hot-spot reference frame (Minster & Jordan 1978)). Relative stress magnitude data in the interior of South America are too sparse to evalutate.

For mid-plate North America (most of intraplate North America, excluding Alaska and all regions west of about the 1000 m contour shown on figures 1 and 2) Richardson & Reding (1991) calculated a 10-fold increase in stress magnitudes across the width of the plate due to drag. They point out that there is no evidence for such a variation; in mid-plate North America the horizontal stresses appear to decrease westward, not increase; deformation in the eastern United States is a mixture of thrust and strike–slip faulting whereas in the central U.S. it is dominantly strike–slip. Driving-drag models would predict an increase in thrust faulting in the central U.S.

A more complete evaluation of drag effects considers actual whole mantle flow patterns. Recent analysis of the geoid and of seismic tomography constrain broad-scale density inhomogeneities within the Earth's mantle. The resultant flow pattern (including the effect of dense masses corresponding to oceanic slabs) has been calculated by Ricard & Vigny (1989). Recently, Bai *et al.* (1991) have computed the stresses in the lithosphere induced by the deep mantle flow pattern predicted by Ricard & Vigny. They superimposed the stresses calculated for this flow model on their proposed non-tectonic reference state of 50–100 MPa deviatoric extension in the continents and compression in the oceans computed from the contrast in density between oceanic and continental lithosphere structures (figure 6*a*). The resultant broad-scale lithospheric stress patterns for the superimposed effects of the mantle flow are shown in figure 6*b*.

One important effect of the stresses due to mantle flow is to put most continents into compression rather than extension (note that the mantle flow stresses include the broad-scale density effects of ridges and slabs). Bai *et al.* (1991) concluded that comparison of their predicted stress states with the observed stress data is premature because the coverage is spotty and so little information on absolute stress magnitudes is available. Furthermore, they state that significant variation in stress patterns are not resolvable for wavelengths less than 5000 km (a distance that roughly coincides to three of their computed grid points).

However, as described in §2 the available stress data define a number of broad areas (on the order of 5000 km on a side) of consistent stress orientations. Several prominent stress patterns predicted by their model poorly match the available stress observations even at this broad 5000 km wavelength. Specifically, Bai *et al.* predict NW–WNW $S_{H,max}$ directions in mid-plate North America which are inconsistent with the observed ENE compression; similarly, strong NE compression predicted for western Europe is in contrast with the observed NW compression. In addition, Bai *et al.* predict enormous deviatoric extensional stresses related to upwelling in the central Pacific, a region noted for its lack of intraplate seismicity. The predicted extensional stresses are greater than 100 MPa, nearly twice as large as predicted anywhere else on the globe, even the seismically active Indian Ocean. However, this prediction for the Pacific was based on an assumption of one-layer convection which also overpredicts the geoid high, it is likely that a two-layer convection model may be more appropriate (C. Froidevaux 1991, personal communication).

4. Conclusions

The mid-plate stress field in both the oceans and continents is largely compressional with one or both of the horizontal stresses is greater than the vertical stress. The orientation of the mid-plate stress field is consistent with plate-motion directions in many areas and seems to be influenced dominantly by plate-boundary forces (locally modulated by bouyancy forces). The geometry of the plate boundaries are extremely important in determining the broad-scale stress orientation patterns.

Regional variations in relative stress magnitude are defined based on direct stress measurements and the style of faulting. In most mid-plate areas a combination of thrust and strike–slip faulting régimes are observed. Most regions of normal faulting occur in areas of high elevation; the extensional stress régime in these areas can be attributed to superimposed bouyancy forces related to thickened crust and/or thin lithosphere and active mantle upwelling. To explain observed local stress rotations, the upper crustal stresses derived from these bouyancy forces must be several times the magnitude of the regional stress differences due to plate-boundary forces. In areas where normal faulting occurs in close proximity to thrust and strike–slip faulting (such as in major mountain belts but also within western Europe) the $S_{H,max}$ orientations are consistent for all three stress régimes.

Evaluating drag is difficult with stress orientations alone. First-order patterns of stress orientation are well explained as the result of plate-boundary forces acting on plate geometry. However, in several cases the observed $S_{H,max}$ directions are subparallel to absolute velocity directions as well. Evidence for lateral stress gradients with an order of magnitude variation in stress values across large plates (predicted for models in which drag dominates (Richardson 1991; Richardson & Reding 1991)), is not observed in the relative stress magnitude data. Similarly, the broadest-scale patterns observed in the stress data do not match predictions of stress orientations related to whole mantle flow inferred from current generation seismic tomography models (Bai *et al.* 1991).

Thus the intraplate lithospheric stress fields appear to be dominated by plate-boundary and local bouyancy forces. However, since lithosphere plates are merely the uppermost part (boundary layer) of the Earth's convection system, the stress fields can always be linked ultimately to mantle convection.

The World Stress Map project is a collaborative effort of about 30 scientists worldwide who have actively compiled and contributed data: J. Adams, M. Assumpcao, S. Bell, E. A. Bergman, W. Bosworth, D. Denham, J. Ding, N. Gay, S. Gregersen, T. N. Gowd, G. Grunthal, A. Gvishiani, K. Jacob, P. Kropotkin, R. Klein, P. Knoll, M. Magee, J. L. Mercier, V. Mount, B. Müller, N. Pavoni, K. Rajendran, S. I. Sherman, O. Stephansson, D. Stromeyer, G. Suarez, M. Suter, A. Udias, Z. H. Xu, M. Zhizhin and M. D. Zoback.

References

Assumpcao, M. 1991 The regional intraplate stress field in South America. *J. geophys. Res.* (In the press.)

Artyushkov, E. V. 1973 Stresses in the lithosphere caused by crustal thickness inhomogeneities. *J. geophys. Res.* **78**, 7675–7708.

Bai, W., Vigny, C., Ricard, Y. & Froidevaux, C. 1991 On the origin of deviatoric stresses in the lithosphere. *J. geophys. Res.* (In the press.)

Chase, C. G. 1979 Asthenospheric counterflow: a kinematic model. *Geophys. Jl R. astr. Soc.* **56**, 1–18.

Cloetingh, S. & Wortel, M. J. R. 1986 Stress in the Indo-Australian plate. *Tectonophys.* **132**, 49–67.

England, P. & Houseman, G. 1989 Extension during continental convergence, with application to the Tibetan. *J. geophys. Res.* **94**, 17561–17579.

Fleitout, L. & Froidevaux, C. 1982 Tectonics and topography for a lithosphere containing density heterogeneities. *Tectonics* **1**, 21–56.

Fleitout, L. & Froidevaux, C. 1983 Tectonic stresses in the lithosphere. *Tectonics* **2**, 315–324.

Frank, F. C. 1972 *Plate tectonics, the analogy with glacier flow and isostasy. Flow and fracture of rocks.* Geophysics Monograph Ser. 16, pp. 285–292. AGU.

Froidevaux, C. & Isacks, B. L. 1984 The mechanical state of stress of the Altiplano-Puna segment of the Andes. *Earth planet. Sci. Lett.* **71**, 305–314.

Grunthal, G. & Stomeyer, D. 1991 The recent crustal stress field in central Europe – trajectories and finite-element modeling. *J. geophys. Res.* (In the press.)

Hager, B. & O'Connell, R. J. 1979 Kinematic models of large-scale flow in the Earth's mantle. *J. geophys. Res.* **84**, 1031–1048.

Henry, W. J., Mechie, J., Maguire, P. K. H., Khan, M. A., Prodehl, C., Keller, G. R. & Patel, J. 1990 A seismic investigation of the Kenya rift valley. *Geophys. J. Int.* **100**, 107–130.

Isacks, B. L. 1988 Uplift of the central Andean plateau and bending of the Bolivian orocline. *J. geophys. Res.* **93**, 3211–3231.

Jurdy, D. M. & Stefanick, M. 1991 Stress observations and driving forces models for the South American plate. *J. geophys. Res.* (In the press.)

Kusznir, N. J. & Bott, M. H. P. 1977 Stress concentration in the upper lithosphere caused by underlying viscoelastic creep. *Tectonophys.* **43**, 247–256.

Kusznir, N. J. & Park, R. G. 1984 Intraplate lithosphere deformation and the strength of the lithosphere. *Geophys. Jl R. astr. Soc.* **79**, 513–538.

Kusznir, N. J. & Park, R. G. 1987 The extensional strength of the continental lithosphere: its dependence on geothermal gradient, and crustal composition and thickness. *Geol. Soc. Lond. Spec. Publ.* **28**, 35–52.

Lister, C. R. B. 1975 Gravitational drive on oceanic plates caused by thermal contraction. *Nature, Lond.* **257**, 663–665.

Meijer, P. Th. & Wortel, M. J. R. 1991 The dynamics of motion of the South American plate. *J. geophys. Res.* (In the press.)

Minster, J. B. & Jordan, T. H. 1978 Present day plate motions. *J. geophys. Res.* **83**, 5331–5354.

Molnar, P. & Tapponnier, P. 1975 Cenozoic tectonics of Asia: effects of a continental collision. *Science, Wash.* **189**, 419–426.

Müller, B., Zoback, M. L., Fuchs, K., Mastin, L., Gregersen, S., Pavoni, N., Stephansson, O. & Ljunggren, C. 1991 Regional patterns of tectonic stress in Europe. *J. geophys. Res.* (In the press.)

Parsons, B. & Richter, F. M. 1980 A relation between driving force and geoid anomaly associated with the mid-ocean ridges. *Earth planet. Sci. Lett.* **51**, 445–450.

Ricard, J. & Vigny, C. 1989 Mantle dynamics with induced plate tectonics. *J. geophys. Res.* **94**, 17543–17559.

Richardson, R. M. 1991 Ridge forces, absolute plate motions, and the intraplate stress field. *J. Geophys.* (In the press.)

Richardson, R. M., Solomon, S. C. & Sleep, N. H. 1979 Tectonic stress in the plates. *Rev. Geophys. Space Phys.* **17**, 981–1019.

Richardson, R. M. & Reding, L. M. 1991 North American plate dynamics. *J. geophys. Res.* (In the press.)

Thompson, G. A. & Zoback, M. L. 1979 Regional geophysics of the Colorado Plateau. *Tectonophys.* **61**, 149–181.

Wiens, D. & Stien, S. 1985 Implications of oceanic intraplate seismicity for plate stresses, driving forces, and rheology. *Tectonophys.* **116**, 143–162.

Zoback, M. D. & Zoback, M. L. 1991 Tectonic stress field of North America and relative plate motions. In *Neotectonics of North America, decade of North American geology* (ed. B. Slemmons), Boulder, Colorado: Geological Society of America. (In the press.)

Zoback, M. L. 1991*a* First and second order patterns of stress in the lithosphere: the World Stress Map project. *J. geophys. Res.* (In the press.)

Zoback, M. L. 1991*b* Stress field constraints on intraplate seismicity in eastern North America. *J. geophys. Res.* (In the press.)

Zoback, M. L., Nishenko, S. R., Richardson, R. M., Hasegawa, H. S. & Zoback, M. D. 1986 Mid-plate stress, deformation, and seismicity. In *The western North Atlantic region, Decade of North American geology* (ed. P. R. Vogt & B. E. Tucholke), vol. M. Boulder, Colorado: Geological Society of America.

Zoback, M. L. & Zoback, M. D. 1989 Tectonic stress field of the continental United States. *Geol. Soc. Am. Mem.* **172**, 523–539.

Zoback, M. L. *et al.* 1989 Global patterns of tectonic stress. *Nature, Lond.* **341**, 291–298.

Zorin, Y. A. & Rogozhina, V. A. 1978 Mechanism of rifting and some features of the deep-seated structure of the Baikal rift zone. *Tectonophys.* **45**, 23–30.